Engineering
Project Management

COST ENGINEERING

A Series of Reference Books and Textbooks

Editor

KENNETH K. HUMPHREYS
American Association of Cost Engineers
Morgantown, West Virginia

Additional Volumes in Preparation

Engineering Project Management

Frederick L. Blanchard

Arabian American Oil Company
Dhahran, Saudi Arabia

Marcel Dekker, Inc. **New York · Basel · Hong Kong**

TA
190
.B56
1990

Library of Congress Cataloging-in-Publication Data

Blanchard, Frederick L.
    Engineering project management/Frederick L. Blanchard.
        p. cm. -- (Cost engineering; 16)
    Includes bibliographical references and index.
    ISBN 0-8247-8119-8 (alk. paper)
    1. Engineering--Management. 2. Industrial project management.
I. Title. II. Series: Cost engineering (Marcel Dekker, Inc.); 16
TA190.B56   1990
658.4'04--dc20   90-3926
                                                                    CIP

This book is printed on acid-free paper.

MARCEL DEKKER, INC.
270 Madison Avenue, New York, New York 10016

Current printing (last digit):
10 9 8 7 6 5 4 3 2 1

PRINTED IN THE UNITED STATES OF AMERICA

Preface

The management of projects presents a unique opportunity for the application of management principles and systems. In a fixed time frame, within an established budget, one executes all the functions of management: planning, organizing, staffing, and controlling. *However* The literature is replete with concepts covering the ideal methods for nearly all management functions. Few acknowledge that the real world is often an unreceptive format for idealized concepts and all must be modified to accommodate the vagaries of any human endeavor. This text, while presenting a compilation of management theory, attempts to aid the practitioner by taking a pragmatic approach to application, dictated by experience. Many of the aspects of project management covered are applicable to all types of projects. Some, however, must be adapted from the engineering and construction setting that provided the bulk of that experience.

The responsibilities of management have been identified as planning, organizing, staffing, and controlling. The format of the text follows these major headings. Another section covers special topics such as megaprojects, which are of particular interest and span the whole of management responsibilities. The intent of each section is to identify the need for the manager to perform each of his major functions, a compilation of methods and techniques that may be applied, and the modifications that might be required in actual application. Every effort must have an objective. It is purposefulness which gives acceleration to effort. A goal may never be reached, but because of its presence, effort can be focused.

The literature codifies historical actions and catalogues them to support the theories, many of which are easily recognized by today's eager practitioners. Many attempts at application of these theories, however, fail. The problem is not the theory itself, but the natural inertia of human systems and their resistance to change. The implementation of theory is a twofold effort: the first, to set up a receptive system; the second, to apply the theory. It is really the first which is the most difficult and should get our primary attention.

A project manager is not going to find the corporate environment totally to his liking. Neither will he find that he can bend that environment to his will. The best that he can hope for is to cope with it as it exists. Fortunately, the environment is usually receptive or accommodating to the proper approach. There are situations, however, in which it may be downright hostile. It is under these conditions that the project manager must function and the success or failure of the effort will be dependent on individual ability to understand the environment and to manipulate it successfully.

The corporate culture of his own company and that of others is the environment in which any individual in project management will find himself. An understanding of these cultures and the relationship of the individual to them is basic to any accomplishment. For this reason I chose to begin the text with this subject. Most successes or failures can be explained by a lack of ability to understand the culture in which any effort is to be undertaken. None of the theories or examples of business practices and techniques in books which you will find in this or any other book is going to be of much use if you don't know how to sell them under the conditions in which you find yourself. If you get nothing more from this text than an understanding of how to sell yourself and your ideas, it will have been worth the effort.

There are, however, few answers here. The text was never conceived to be a cookbook where one could find the recipe to answer all of one's questions. Business is too complex and diverse and any individual experience too limited to provide a prescription for all ills. My objective is to put forward a distillation of the results of the scholarly thought of others as it applies to the management of projects and mostly to the management of people. This is counterbalanced from the perspective of my own experience and observation, whether in concert with or in opposition to current management theory. The goal of the text is to identify the most effective ways to implement the management of projects. This acknowledges that implementing a project is to cause change. To effect change means introducing new ideas and innovation. Within that framework are the ideal methods and, in particular, what *can* be done in contrast to what *might* be done. The practical pressures of project management provide little forgiveness for the procrastinator and little or no opportunity for the crusader.

Although my own experience has been limited to capital projects in the engineering and construction industries, many of the principles and experiences discussed apply to a broad range of projects. The development of a consumer product from laboratory curiosity to mass production can be a project. Replacement of a major machine tool in a factory can be a project. Paper reduction in the corporate office can be a project. For purposes of this text a project is defined as any effort that meets the following criteria:

1. A fixed objective(s)
2. A budget
3. A set time for achievement of the objective(s)

Uncertainties in the world economy have put a damper on the highly publicized, multibillion-dollar projects —known as "mega-projects" —of the last decade. Disasters, such as that which occurred in February 1986 with the Challenger space shuttle, have focused attention on the negative potential of project management. At the same time, interest in the Middle East and North Sea has waned with the general sluggishness of the world economy. These developments overshadow the more numerous, but equally important areas of new product development and increased emphasis on innovation and entrepreneurship, which will provide more than adequate challenge for the project managers of the future. Before the twentieth century ends, the cross-channel link between England and France and an orbiting space station may become reality and gain a deserved share of public attention. Thousands of other projects, important in their own right to those whom they impact, will also be undertaken. It is for the students of project management and those who aspire to become project managers that this text is intended. If it can provide some new idea or reinforcement to the seasoned project manager, so much the better.

This effort encapsulates over 30 years of involvement in the environment of project management. Its ideas and experiences have been molded in great part by the mentors with whom I have had the opportunity to work and to whom I shall always be indebted. The first was Joe Watkins, who initiated me into the real world of construction. John Patterson, with whom I had the pleasure to work in two stages of my career, taught me the benefits of management by exception. These were further refined by John Donald, who rescued me when I was losing confidence. Frank Fugate, of ARAMCO, had the confidence to give me the challenge most project managers can only dream about, projects of vast scope and responsibility. It was he and my last mentor, Dan Barbee, who made possible the opportunity to embark on an entirely new career, after having achieved the ultimate from the first.

It was a wise executive who said that one gets ahead by being pulled up from above and pushed up from below. I have given due credit to those who have pulled me up the ladder of achievement, but I cannot forget the many hundreds to whom I am equally indebted and without whom many of the things for which I am most proud could not have been accomplished. Past associates Jack Fleming, Rudy Bergfield, Hans van Battum, Ray Russum, and Tom Haney were kind enough to review the text and offered valuable comments based on their vast project management experience. Finally, my wife, Barbara, has put up with the constant moving and separations and borne the brunt of my frustrations when things were not going well. It is to all of them that I am grateful and to whom I dedicate this book.

I apologize in advance to the growing number of women in the engineering and construction industries for using only masculine pronouns in the text. It is a habit developed from never having the pleasure and benefit of working with or for women in my long career. As some of my current female students enter these fields, the balance may change and I with it.

Frederick L. Blanchard

Contents

Contents

Engineering
Project Management

1
Corporate Culture

PROLOGUE

Project management is the art of the possible. Possibility can only occur when an idea enjoys the support of those who will provide the necessary resources to bring it to fruition. A risky notion will fall on deaf ears in an organization that is averse to risk. The promoter may even be considered a maverick. An innovative scheme may, however, receive nourishment and the innovator reward and recognition in the enterprising firm. Like the successful gambler who knows when to hold, when to raise, and when to fold up, the successful project manager knows what can be accomplished. By assessing his company, its management, and himself, he knows how and to whom to promote the ideas and actions necessary to complete a project successfully. Corporate culture is nothing more than the personality of the organization as reflected in the words and deeds of its management. Assessing the compatibility of characteristics of your own and the organization's personality will yield valuable insight into which ideas will enlist support and those which won't. It will also enable you to determine the best possible way to promote them.

What is a student, or practitioner, of the management process searching for? Is it a stimulus to generate ideas, or comfort in confirming those already developed? Is it an attempt to acquire and promote a compendium of general thought, or a biased distillation? Whatever the purpose, to be successful his trials need to be attuned to the corporate culture. Trial, in turn, comes only in a receptive culture. To promote ideas, or biases, it is essential to understand the corporate culture, to learn how the company relates to its employees and the outside world, and vice versa. With such knowledge the corporate culture can be understood. Such knowledge will help sell ideas and influence change, and will enhance the executive's ability to manage change, the only business certainty.

Corporate culture, or personality, as used to describe the basis of a company's inner workings is not new. It is not confined to the large corporation and is as old as business itself. It reflects the values of its owners and leaders and provides a framework within which to operate. It is a topic fundamental to the understanding of the management process.

1.0   CULTURAL BASES

A company mirrors the values of its founders, owners, and hired managers. The agglomeration of these values —theoretical, economic, social, human, political, and moral —is called its "value set." It is the fundamental basis by which a company establishes the tenets of its behavior. If its values and the products or services that reflect them are acceptable to the public and responsive to its competition, the company will prosper. If not, a change may reverse the company's fortune or it will eventually vanish. As the company develops, a slow transformation occurs when hired management begins to replace the original owners. Although they are often selected for their similar values, subtle changes are inevitable as these new managers begin to exert themselves. Some old values will linger or be perpetuated depending on their acceptance by the new management. Often they begin to take on a life of their own and become part of the corporate lore. With each change in management the "value set" changes by more or less emphasis on one or more of its constituent values.

> The rising dominants in an effort to placate the declining dominants, may be willing to muddy or obscure their own position to avoid too direct a confrontation of values. Old values lag on and are not easily dislodged from the minds of people. Some aspects of the traditional values may not be particularly important or vexatious to the new dominants, so that they are content to leave them in place. [1]

>   Neil W. Chamberlain

1.1   MEANING TO THE ORGANIZATION

A business is a social organization. It is made up of people who must work together. They must be compatible, share similar values, and be supportive. The status quo is perpetuated by those who do the hiring, particularly the original owners. They select those much like themselves, with whom they can work, or at least who share their values. When the chief steps down, his faithful number two moves into place. Those who reach the pinnacles of power have usually been carefully selected, groomed, and, more important, tested, to ensure compatibility with their selectors. The executive selection process provides continuity and stability in an organization and can result in continued success, if its value set still responds to its environments. It can present a serious problem if changing conditions demand a significantly different value set. Nowhere was this more evident and the impact as traumatic than with the break-up of the American Telephone and Telegraph Co. Bell System [2]. The change from a regulated monopoly to several businesses in highly competitive markets found its value set totally out of tune with its newly acquired competition. Outside executive talent was brought in to infuse new values. Personnel with market-oriented values were hired to catalyze the change. Many employees could not make the change and left voluntarily or through early retirement. Many more struggled on, groping to adapt.

Who recognizes the need to change the basic value set? The answer lies in the degree of change necessary and responsiveness to change. Most organizations maintain a value set that is inherently responsive to changes in its internal as well as external environment. Its leadership is alert to recognize the need for change and flexible to respond. They "stay close to the customer" and have a "bias for action" [3]. Others have directors or lenders who recognize the need for change, prod current management to initiate it, or bring in new management to generate it. Still others have grown lethargic and incapable of adjusting. These become candidates for outside predators, decay to insignificance, or vanish.

The value set is like a flywheel; the bigger it is, the more difficult it is to alter its course. The larger the corporation, the greater number of people who have an interest in maintaining the status quo. If change is needed and is not something that is constantly going on, the natural fears of change are tougher to overcome.

Change will occur naturally through the ever-varying needs of the public or it will be forced by the pressures of the competition. Success, therefore, is measured by a company's ability to manage change. Change is affected by the adoption of new ideas through innovation. Innovation thrives in an environment of fertile minds, nourished by enlightened management. In cloning itself, management cannot lose sight of dissenting views as a means to test and temper prevailing thought and provide a solid foundation for the adjustments demanded by change.

1.2   MEANING TO THE INDIVIDUAL

Performance counts; achievement counts. At lower levels
in the organization, it may be all that counts. But as you
move up through the management hierarchy, how well you
fit in with the organizational culture becomes increasingly
important. Eventually (early in some and later in others),
you must reach a point at which the finest performance in
the world will not move you ahead unless you are also seen
as "one of us" by the more powerful managers. [4]

Ellen J. Wallach

Corporate culture has significant meaning for its employees.
Compatibility will result in synergy. Incompatibility will result in
delusion, disappointment, depression, and ultimately separation.
This is true irrespective of who made the judgment, the employee or
the employer. The individual must make an assessment of himself
and how he fits in the culture. The company must assess itself and
how it is perceived by its employees. Individual behavior is a reflection of the values developed during childhood and adolescence.
By the time of entry into the job market, attitudes toward work and
people are essentially established. This forms the individual's value
set. Work can be rewarding or a drag and association with others
to be cultivated or left alone. There are infinite positions between
these extremes and innumerable combinations between them and
other values in the value set. Each individual is just that —a
unique, complex mixture of which, like snowflakes, there are no two
alike. Understanding or modifying the behavioral expression of the
value set may be sought through T-groups, sensitivity training, or
Managerial Grids [5]. These are various mechanisms that attempt to
provide the individual with an understanding of how he is perceived
by others. Very little, short of therapy, will result in changing it.
The best expectation is to maintain subdued those behavioral expres-
sions considered unacceptable or detrimental to the company, or
project objective.

1.3   THE CULTURAL INVENTORY

An inventory of the value set will enable a better understanding of
the individual and the company and a comparison should shed some
light on how variations can be interpreted and used. Vernon,
Allport, and Lindzey [6] developed a "Study of Values" from work originally suggested by E. Spranger [6a]. Their classification of values, with minor variations, provides a good starting point for the
inventory.

Value	Manifestation
Theoretical	Interest in the discovery of truth
	Approach to knowledge
Economic	Use of the means of production distribution and consumption
Social	Appreciation or response to the nature of the non-materialistic aspect of things
Human	Orientation to people
Political	Means of exerting influence and control
Moral	Attitudes, beliefs, customs

How these values manifest themselves in actual deeds vary with companies as they do in individuals. For example, in the theoretical area there are those, like the Bell Laboratories of American Telephone and Telegraph Co., which explore the aspects of pure research. The work may have no immediate practical application or investment return. At the other end of the spectrum are companies that do no research at all and copy that of others.

More emphasis is being placed on the area of a company's social responsibility. Some, like Johnson & Johnson, which voluntarily pulled $100 million worth of Tylenol off the market in a 1985 product-tampering scare, are at the high end of the spectrum. They place significant value on their reputation within the community. Their actions were based on a policy credo set down over 40 years ago by one of the founders, General Robert Wood Johnson. He wrote,

> Institutions, both public and private, exist because the peo-people want them, believe in them or at least are willing to tolerate them. The day has passed when business was a private matter —if it ever really was. In a business society, every act of business has social consequences and may arouse public interest. Every time business hires, builds, sells or buys, it is acting for the . . . people as well as for itself, and it must be prepared to accept full responsibility. [7]

Others, such as Hooker Chemical Co. in the case of pollution of the Loveland Canal and Johns Manville in the asbestosis case, represent another aspect of social responsibility. At fault or not, they are

striving to recover from situations that are the probable result of
neglect of this aspect of a company's relationship with its public.

The issue here is not what is good or bad, or what is expected
by today's standards. Neither is it a question of whether one's
actions are profitable or unprofitable. High and low profits, or no
profits at all, may exist simultaneously in different companies at
both ends of the spectrum. The issue is how a company's actions
reflect its values and how those values match those of the individual
employee and the public. Each of these values has a different importance with respect to the others and over time.

Without being exhaustive, each of these values will be examined
and how they may relate to each other.

1.3.1   Theoretical Values

One should not confuse a theoretical approach with simply a penchant for research. A search for the truth, in particular when it
appears to have no commercial value, is laudible. On the other
hand, many have experienced the consequences of paralysis by
analysis. The time for action has passed and those who cannot act
request another study.

There are those who prefer the mental rigors of investigation.
They gravitate toward organizations whose survival depends to a
great extent on discovery and development. Discovery and development must be balanced with a knowledge of what the market demands, how to produce what is developed, and how to sell it. Failing this, opportunities for further development will diminish or dis-
appear altogether. In management, including that of research, one
must sense when to fish or cut bait. Progress will slow if those
who must have every question answered prevail. As one moves up
the organization ladder, it becomes apparent that decisions are large-
ly made on the basis of evaluation of the conviction, experience,
and confidence of one's subordinates. Although there may at one
time have been sufficient knowledge to sift and judge the facts, it
has become dulled. The real job is to be able to judge others'
presentation of the facts, their conviction and thoroughness in developing them.

Individuals and organizations run the gamut from flying by the
seat of their pants to paralysis by analysis. A judgment of where
one and his company fits is important from many aspects. As an
individual, flip decision making may meet with instant success, but
more probably instant failure. It is not successful in the long term.
Undue deliberation in pursuit of more information may result in
missed opportunities. An organization at either of these extremes
will not likely fare well. One with a penchant for quick decisions
in an organization whose mode tends to analysis in depth is likely

to become frustrated. Insecurity will most certainly result if an individual who tends toward thorough analysis finds himself with a management that prefers hip shooting.

The February 1986 explosion of the American space shuttle Challenger presents an excellent case to examine theoretical values. The National Aeronautics and Space Administration (NASA) had achieved an excellent reputation for its strength of technical achievement and, with only one previous exception, an excellent record of safety. The committee that investigated the disaster, however, castigated the NASA management and its procedures for allowing expediency to outweight prudence.

The cause of the disaster was declared to be a faulty seal in the booster rocket. This was exacerbated by the low temperature conditions prevalent at the time of the Challenger launch. The investigation brought to light the fact there had been prior problems with this same seal. It had also been reported in great detail and with expressed concern. Before the Challenger launch, engineers for the booster rocket's manufacturer, Morton Thiocol, had strongly recommended the launch be aborted. They were later overruled by their management. Their conviction and courage in coming forward during the early stages of the investigation had a significant impact on the direction of the investigation and probably the future of NASA.

Whether the situation that existed at NASA in February 1986 evolved from values that were initially different or whether the earlier record was pure chance may never be known. Political pressure, economic pressure, or a combination of both caused the evolution of a management style that condoned a higher level of risk than was ultimately deemed acceptable. An engineer who requires all the answers is averse to risk. If he works under risk-taking decision makers and pressure of deadlines, he is in a conflict situation. Pressing for additional time to search out the answers or requesting additional funds would likely result in negative responses and finally frustration.

1.3.2   Economic Values

The industrial revolution and mass production have all but eliminated the artisan form of production that preceded it. Along with it went identification of individual accomplishment. One has now only a small piece of a greater whole to call his own. Some have recognized the negative consequences of this change. It becomes reflected in recognition of the contribution of the individual and the quality of product.

At the opposite end are those who place greater value on the ends as opposed to the means. The individual recognizes work as a

necessary evil to obtain the things he really values. The employer
considers the worker as a resource to be exploited. The product or
service is a means to profit. The price will be as high as the
traffic will bear.

Before the industrial revolution many product production efforts
could be traced directly to the individual. Work or labor was an
individual pursuit. Its results were identified with or traceable to
the artisan, craftsman, or laborer who performed it. Success was a
measure of one's skill and pride its nonmaterial reward. Many companies were founded on the basis of producing quality which is reflected in the pride of the individual employee in the product or
service. The customer recognizes this and is willing to pay a
premium.

Many may recall the general opinion held of Japanese products
before and just after World War II. It was one of poor quality and
a lack of originality. The commonly held picture of the Japanese
engineer was of a camera carrier whose appetite for information was
voracious. The consequence was the appearance of a cheap imita-
tion. In an unprecedented national campaign to reverse this image,
the Japanese have succeeded in a wholesale modification of their own
attitude. The key to their success was recognition that a change
in values was what was required. They now have a worldwide reputation for high-quality products.

1.3.3   Social Values

The past few decades have seen a marked increase in awareness of
the quality of life. This has been manifest in the volume of law-
suits in attempts to preserve the environment, demonstrations against
nuclear power, and the increase in membership and activity of organizations such as the Sierra Club and Greenpeace.

Moral and political values provide the order necessary for survival. Theoretical, human, and economic values suffice to quantify
our needs as individuals. Social values reflect our communion with
our fellow man. Certainly, in the developed countries, preservation
for the majority is no longer an issue. Also, personal values are
reflected in the systems that have been established. As needs are
satisfied reflecting these values, desire to implement a higher set of
values can be addressed. These take the form of improvement in
the quality of work life, preservation of the environment, and a
general increase in the appreciation of the arts.

Social responsibility activists are in the spotlight mainly because
of the methods chosen to put forward their ideas. Picketing, boycotts, and legal action, aided by a sympathetic press, have increased
and attracted larger numbers of supporters. There are few business executives among this highly vocal minority. The primary

reason for this is that few activists of any persuasion ever reach
the top echelons of corporate life. If they do, they are aware more
will be accomplished by persistent persuasion than activist blud-
geoning.

The vocal minority are a necessary bellwether of issues that
need to be addressed. Like the first symptoms of a disease, they
are the harbingers. Hazardous waste disposal, nuclear power,
famine in Africa, apartheid, and acid rain are major issues. Many
of these issues have been with us for some time. Emphasis ebbs
and flows, altered by events. There was the Union Carbide gas
plant leak in Bhopal, India, in 1985, which took over 2000 lives,
and the 1986 Chernobyl nuclear disaster, whose real toll may not be
known for years. Hundreds of companies are involved in these issues. Hundreds more will undoubtedly be added to their number
before these problems go away. One thing is, however, certain —
they will be replaced by other problems. As difficult as some of
these problems will be to solve, so too is changing the corporate
culture. This change may be measured in years, but can take considerably less time if change is initiated from within rather than
without.

Most organizations are conservative. It would do well to remember that, unless prodded by strong outside forces, they are
lethargic in making appropriate changes. An old saying in the machinery business — "if it runs, leave it alone" — applies in most
cases. This does not mean that old ideas should be ignored or
principles compromised. More will be accomplished with a gentle
nudge than a hard push.

1.3.4   Human Values

This value has the most recognizable characteristics and has been
the most extensively studied. Its impact on the interrelationship
between the individual and the company also makes it one of the
most important. There are four aspects to this interrelationship:
the individual himself and the individual with respect to his supervisor, his own subordinates, and the company.

How this value manifests itself is called style. In relations to
others, individuals vary from loners to joiners, from autocratic to
democratic. Likewise, the company and others can be classified in
relation to the individual in a similar fashion.

Observable style is a reflection of personality. As such, it is
unlikely to be altered significantly by situational change. The true
personality is extremely difficult to suppress. The range of style
is broad, from the isolation of the hermit to those who are insecure
unless part of a group. In the main, people can work alone and
take pride in individual accomplishment and recognition. At the

same time, they can contribute unselfishly to group enterprise, enjoy the companionship and camaraderie of a group, and share in its successes or failures. Over time and many varied situations, individual personality can be categorized.

Autocratic or democratic describes the style of the relationship with others. These styles can be classified into four distinct groups [8]:

Autocratic —Gathers and analyzes the facts and makes a decision.
Consultative — Gathers and analyzes the facts, tests the decision on others before making it.
Participative — Defines the problem and seeks advice of the group and they form a collective decision.
Democratic —Defines the problem to the group, charges them to make a decision, and supports the decision when made.

Some theorists suggest that these styles can vary in individuals depending on the requirements of the situation [9]. I believe that a person who can be classified as democratic could conceivably adopt another style as the situation demands, but a person who is autocratic would find such change extremely difficult. There are also situations and professions that demand certain styles. For example, the military and police and fire departments operate in situations where democratic decision making or problem solving could result in catastrophe. Selection processes and training programs for these professions attempt to select and develop persons who will accept autocratic rule.

1.3.5   Political Values

Power can be classified as intrinsic or extrinsic. Intrinsic power is power that evolves from the legal and positional base of the company and those who manage it. It has been alternatively referred to as legitimate power. Extrinsic power is power that has been yielded by those who can then be controlled. It is difficult to separate the political and social values. How a person or organization exerts its influence and power is manifest in the style in which they are classified. Then too, autocratic style is often confused with predominant utilization of intrinsic power and democratic style with extrinsic power.

Corporate style in dealing with people varies. The Japanese have turned consensus management into an art form. The average Arab businessman would no more consult his subordinates on a decision than do a deal without bargaining. The meanings to be learned from this go a long way in explaining why most Americans and many Europeans, have difficulty in doing business outside their own geographic area. The majority of Western businesses function in a style somewhere in the middle of these extremes.   The style of individuals is, however, of primary concern.

"The Japanese decision making is neither top-down nor bottom-up.   It is 'MUD' — 'middle up and down' Cheuk C. Kwan, Yoshikuni Takuhashi [10].   Whether supervising others or simply working and dealing with people in a peer relationship, it is done with a certain style.   This may vary from situation to situation, but in the majority of cases, action is in a consistent manner somewhere in the continuum between autocracy and democracy.   How an individual relates to others depends to a great extent on the compatibility of the individual styles.

The press relates many instances of a senior executive or key individual leaving a company because his style was not compatible with what he thought the company wanted.   Even the illustrious Peters and Waterman, authors of the best seller *In Search of Excellence*, left their employer, McKinsey & Co.   Waterman was later quoted as saying, "I didn't like doing my thing in an environment that wasn't sure it was the right thing to do" [11].   These are only the top people and the tip of the iceberg.   Beneath the surface are literally thousands whose moves from one company to another are important only to themselves.   There are additional millions who labor begrudgingly because they cannot or do not want to change.

There are many reasons for individual moves or unhappiness in one's current job.   The reasons are complex, but a predominant factor is the incompatibility of style and personality.   The greater the differences, the quicker the separation or the more frustrating the relationship.   It is important to determine individual styles and compatibility with them.   On this compatibility will depend the pleasure or frustration of the work experience and where the future relationships might lead.

Corporate leaders and managers do get to pick those who move up through the organization.   On initial job entry, personality and style factors on both sides are less well known and least considered. They increase in importance as one moves up the ladder.   Without performance one hardly gets noticed.   That performance will go unrewarded without compatibility with the style and personality the management expects.   The nonconformist, whose performance cannot help but be noticed, will ultimately find himself topped out at a certain point.   His nonconformity is more of a threat to the culture than are the benefits of his performance.

The first supervisory positions one attains result primarily from display of expertise in one's respective field.   In the supervision of others, performance begins to depend more and more on what others can and will do.   Knowledge becomes more generalized and less specific.   Moving up successfully depends to a large extent on how well others can be induced to work together.   Doing this means

picking people who can work together without rubbing each other
the wrong way or keeping conflict to the minimum. As one rises
through the organization, it becomes apparent that the pickers at
the higher levels pick the pickers at the lower levels. At each
level, the individuals become better known and the chances of picking an incompatible person grow progressively less. This has a
significant meaning for both the individual and the company.

Several choices are open in pursuit of ambition. Organizations
can be sought out which are compatible in style or adjustments can
be made to reduce or minimize any differences. Getting ahead
would then depend to a great extent on performance and less on
style. The focus should be on the supervisor and where he stands
in relation to his superior and the company as a whole. If you
can't get on with your boss, it is the unusual case that your name
is going to be brought forward for any promotional opportunity.
The best one could hope for in this situation is that the boss is
just as unhappy with the relationship and will move you on to someone else. The stigma of incompatibility will probably move along
with you. There is a good chance that if your name comes up again,
your old boss will be around to add his comment. It is not likely to
be favorable. Under these circumstances, if one cannot change, it
is probably well to consider moving on, or reconcile oneself to having
reached his peak.

Compatibility is important to advancement. The alternative to
seeking out a company where one's style is accepted is to modify
one's style, or subdue it. Whether this change is permanent or not,
the main objective is to be constantly aware of it and its importance
not only in getting ahead, but in many cases to getting the job done.
The people who work with you and particularly those who work for
you should be trying to do the same thing. The company itself may
provide help and encouragement in this effort with development programs such as the "Managerial Grid" [12], developed by Blake and
Mouton. The Grid is shown in Figure 1.1.

This system provides for determination of a manager's assumptions and values regarding managing people. A series of exercises —
individual, as well as group —are conducted in order for managers
to compare their style with the 9,9 orientation. It is presupposed
that the objective is to move to a 9,9 managerial style, which inte-
grates a high concern for both company objectives and people objectives. This is accomplished by improving communications and
understanding between individuals and groups. Improvement in communication and understanding between people of opposing views is
beneficial; however, the system does have its critics. Since it is
usually the company or consultants whom the management pick who
administer such programs, it is not difficult to imagine that hoped-
for results are clones of the current thinking. Even Blake and
Mouton admit,

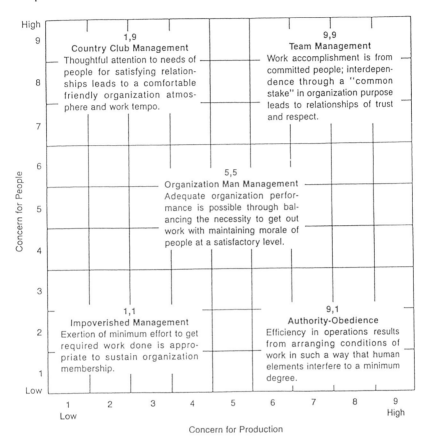

High
9 | 1,9
Country Club Management
Thoughtful attention to needs of people for satisfying relationships leads to a comfortable friendly organization atmosphere and work tempo. | 9,9
Team Management
Work accomplishment is from committed people; interdependence through a "common stake" in organization purpose leads to relationships of trust and respect.

5,5
Organization Man Management
Adequate organization performance is possible through balancing the necessity to get out work with maintaining morale of people at a satisfactory level.

1,1
Impoverished Management
Exertion of minimum effort to get required work done is appropriate to sustain organization membership. | 9,1
Authority-Obedience
Efficiency in operations results from arranging conditions of work in such a way that human elements interfere to a minimum degree.

Concern for People (vertical axis, High 9 ... Low)

Concern for Production (horizontal axis, 1 Low 2 3 4 5 6 7 8 9 High)

Fig. 1.1   The Managerial Grid.   (From Robert R. Blake and Jane Srygley Mouton. *The Managerial Grid III: The Key to Leadership Excellence.*   Gulf Publishing Company, Houston, Copyright © 1985, p. 12.   Reproduced by permission.)

while it is often tempting to see the fundamental problem of organization as inherent in the individuals employed or in the structure of the organizational system, it is becoming more apparent that the way individuals behave and the way the structure is arranged—reflects the culture of traditions, precedents, and past practices that has grown, like a mound of consequences from the past. [13]

Recognition of the culture that exists is important from the standpoint of its entrenchment and difficulty in effecting change. If the company is responsive to its environment and is successful, it can be assumed that its culture fits its needs. If it is not, self-examination may reveal that its culture is out of synchronization with its environment and requires change. All too often corporate management is blind to its own shortcomings and the result is relegation to the fringes of marginal existence or upheaval from the outside. When major changes appear necessary, it is often outsiders, who have not been hypnotized by the existing culture, who bring them about. Management should maintain a balanced culture which encourages "positive conflict." This is an opposition in ideas and style to instill vitality, which combine positively as opposed to negatively. Bringing about positive conflict is not easy. It takes a determined effort and routine vigilance. There are several steps that will help.

1. Establish through policy and top management action that no procedure or standard is inviolate. All can be changed if a better way is available.
2. Carefully monitor training programs to ensure that cloning is not the result. Programs for development and promotion should include profiling of candidates and reviews to ensure balancing of opportunity.
3. Decision making should be such that other viewpoints are sought but responsibility is clear.

The measure of the bureaucracy is the thickness of the "General Instructions" or the "Procedures Manual." They are a legacy of the scientific management principals enumerated by Frederick W. Taylor. There is no better way to ensure that those you lead will act just like you do than to leave behind a clear set of instructions as how to act and what to do in every conceivable situation. The end result will be the stifling of initiative, slow response to change, frustration, and ultimately departure of the best personnel. It is a direct route to stagnation. There are, of course, good reasons to establish some rules which provide a routine enabling people to follow what has been done in the past. Beyond that, the guidelines for business should be the corporate goals and objectives controlled only by the budget and time targets.

Training and development programs as well as replacement plans for management positions should be based on a policy of encouraging equal opportunity for personnel of varying styles. This means selecting or designing training programs emphasizing results from varying approaches. Programs selected should have specific mechanisms for measuring results in order to evaluate them. A formal training organization should have objectives and be measured by results just like any other organization. Personal profiles should be maintained on all personnel and updated routinely. These profiles should include a measure of knowledge, skills, and personality.

Selection of personnel for development and promotion should get
high-level review to ensure balance in style.

> The people who share a bureaucratic culture learn their
> repeatable tasks, how to behave towards each other, what
> to expect from each other, and what each other's specific
> tasks are. They also figure out the true reward system in
> the organization and how to advance in the departmental
> structure. The organization supports this behavior with
> rewards for conformity. [14]

Robert J. Graham

If the thickness of its procedures manual represents the heart
of the bureaucracy, consensus management is its soul. If the procedures don't provide the answer to every question, requiring all
to agree will guarantee its sterility and perpetuate their intent.
The search for opportunities and the solution of problems should be
conducted in an uninhibited environment. Some of the best ideas
and solutions to problems have come from "left field." Some control
is essential to ensure timely choice is made. This is accomplished
by making an individual responsible for it. The individual is free
to adopt and exercise his own style, be it autocratic or democratic.
Accountability will enhance seeking out the best advice, yet bring
about a timely result.

1.3.6 Moral Values

Moral values are heavily influenced by religious upbringing and the
customs observed during development. What constitutes acceptable
and unacceptable conduct is of infinite variety. Despite religious,
cultural, and geographic differences, there is still significant uni-
versality. Society attempts to define what is acceptable and to reg-
ulate behavior through laws, suasion, and occasionally censorship
and suppression.

People act in a certain fashion because they believe it is the way
to act in order to attain their individual objectives. Society provides
incentives or disincentives for those actions. The continuum of
values then becomes an individual measure with respect to those of
society. Values will change in both individuals and society over
time and societies themselves differ in their values. What is acceptable in one society may be unacceptable in others. The individual must assess the consequences of deviation from the accepted
norm.

Moral values are deeply rooted. The first things learned are
what is acceptable and unacceptable conduct. These concepts of
good and bad are carried over into business life. The laws, rules,
and resulations that control conduct reflect the collective morality

of the populous as a whole. These controls still permit a wide range of individual and corporate conduct. Large businesses are composites of many moral inputs and, because they are exposed in the public fishbowl, are the most neutral. In smaller businesses, particularly those that are owner controlled and managed, it is easier to identify differences between an individual and the company. The moral values of the company are more likely to reflect those of the owner and are less complex.

Individual and corporate ethics has received considerable attention as a consequence of activities on Wall Street and revelations of alleged overcharging on government contracts by defense industry contractors. These have been followed by demands for ethics courses in our business schools and a general return to fair conduct. Some would say this is the result of a relaxation of our moral attitudes, but these things went on before. The robber barons and child labor exploiters of the early twentieth century were certainly no saints. The difference is that time has given more and more opportunity to be tempted toward avarice and greed. If the early establishment of a value set is to be believed, morality must be learned at an early age in the family, the schools, and the church. The communications media must also take a responsibility in not promoting the unethical as heroes.

1.4   THE VALUE ASSESSMENT

Ellen Wallach has developed a cultural inventory which, with some elaboration, provides a good basis for assessing the type, style, and personality of an organization.* Matching the assessment categories above with those of Vernon, Allport, and Lindzey provides a basis to determine where an organization and an individual fit. This is not to say that an individual's perception of himself and of his employer will match that of another. It is, however, irrelevant, because it is the individual's perception which ultimately matters.

   Theoretical — Risk Taking (a), Creative (g)
   Economic — Results-Oriented (f), Pressurized (k), Enterprising
       (s), Cautious (u), Driving (w)
   Social — Collaborative (b), Hierarchical (c), Structured (j),
       Established, Solid (t)

---

*Ellen J. Wallach, Individuals and organizations: The cultural match, *Training and Development Journal*, February 1983, pp. 29–36. Reprinted by permission of the publishers.

	Describes (me and)[a] my organization			
	Not at all	A little	A fair amount	Most of the time
	0	1	2	3
a. Risk Taking	_____	_____	_____	_____
b. Collaborative	_____	_____	_____	_____
c. Hierarchical	_____	_____	_____	_____
d. Procedural	_____	_____	_____	_____
e. Relationships-Oriented	_____	_____	_____	_____
f. Results-Oriented	_____	_____	_____	_____
g. Creative	_____	_____	_____	_____
h. Encouraging	_____	_____	_____	_____
i. Sociable	_____	_____	_____	_____
j. Structured	_____	_____	_____	_____
k. Pressurized	_____	_____	_____	_____
l. Ordered	_____	_____	_____	_____
m. Stimulating	_____	_____	_____	_____
n. Regulated	_____	_____	_____	_____
o. Personal Freedom	_____	_____	_____	_____
p. Equitable	_____	_____	_____	_____
q. Safe	_____	_____	_____	_____
r. Challenging	_____	_____	_____	_____
s. Enterprising	_____	_____	_____	_____
t. Established, Solid	_____	_____	_____	_____
u. Cautious	_____	_____	_____	_____
v. Trusting	_____	_____	_____	_____
w. Driving	_____	_____	_____	_____
x. Power-Oriented	_____	_____	_____	_____

[a]Author addition.

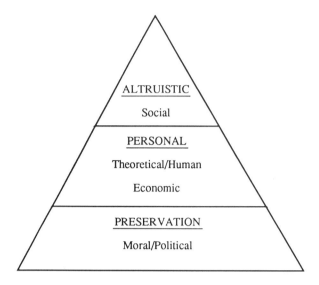

Fig. 1.2   The value hierarchy.

Human — Collaborative (b), Relationships-Oriented (e), En-
    couraging (h), Sociable (i), Stimulating (m), Personal
    Freedom (o), Equitable (p), Safe (q), Challenging (r),
    Established, Solid (s), Trusting (v)
Political — Hierarchical (c), Procedural (d), Ordered (l),
    Regulated (n), Power-Oriented (x)
Moral — Equitable (p), Established, Solid (t)

The following elaboration should enable the individual to rate
both the organization and himself to determine the match.

Risk Taking — Pushing the limits of knowledge and opportunity.
    A willingness to venture when the outcome is uncertain and
    where the rewards and penalties are greatest.
Collaborative — Working together to achieve common goals.   A
    mentor orientation.   Democratic decision making.   A pref-
    erence for group action.
Hierarchical — Established pecking order in form and substance.
    A functional chain of command.   Knowing where one stands
    in regard to position, authority, and responsibility.
Procedural — Codified methods of operation.   Formalized system
    for action and process based on past experience and prac-
    tice.   The level personal activity is prescribed.

Relationships-Oriented — The level at which the organization fosters interpersonal and interdepartmental collaboration. The need or desire to establish contacts to promote individual and group objectives.

Results –Oriented — The measurement of success or failure by outcome as opposed to the methods used to achieve it.

Creative — A penchant to attempt the new and untried, unfettered by what has been done before or how it was accomplished. A willingness to experiment.

Encouraging — Helpful and supportive in a demonstrative fashion.

Sociable — Extending work relationships outside the business environment.

Structured — Form and process follow an established regime. Uncertainty eliminated or minimized by working in and through procedures and hierarchy.

Pressurized — Tight deadlines, high uncertainty of outcomes.

Ordered — Systematic and organized. Outcomes and processes, both individual and corporate, highly predictable.

Stimulating — An environment that excites and inspires the individual employee to excel. The desire to create such an environment as a means to enable subordinates to obtain greater satisfaction and reward for their efforts.

Regulated — Employing structure and procedure for the purpose of control.

Personal Freedom — Management by exception, loose controls, limited restrictions over action and process.

Equitable — Application of rules, procedures, rewards, and punishment in a similar manner to all irrespective of position or personal factors.

Safe — Predictable. Security through the removal of uncertainty.

Challenging — Inspiring the individual to reach beyond current capability to grow in knowledge, experience, and status.

Enterprising — Willingness to employ new methods, adopt or adapt different ideas and attitudes toward the achievement of goals and objectives.

Established, Solid — A sense of security created through long application of effective processes within the capacity and capability of the organization or individual.

Cautious — Unwillingness to test new theories or processes incurring risk.

Trusting — A willingness to allow execution of an assignment with the confidence that it will be carried out in conformance to established procedures and processes without imposing them as formal feedback control.

Driving — The establishment of stressful time and performance
goals. The assumption that all should perform to the
standard of the best.

Power -Oriented — The use of position alone to influence behavior.

Companies have long recognized the importance of fit. The system of selection and promotion perpetuates the values and style of
those who control the system. The individual should recognize the
importance of this fit and be prepared to act accordingly. In the
long term, both must recognize the possibility of change. The company must assure its options remain open to respond to the need
for change and be prepared for it. The individual must sense the
change and the damands for adjustment.

The job interview is the first exposure to the importance of fit.
Some companies take exceptional care to ensure that the candidate
to whom they offer employment will fit their corporate culture.
The primary filters are the choice of schools from which candidates
are drawn, the specific academic disciplines being sought, and the
standard for consideration. The next is an attempt to discover the
personality type through examination of other factors in the candidate's background. These include the types of extracurricular activities which reveal social and political orientation. Elective offices
held are a measure of leadership and management potential. A pro-
file of the candidate can be developed from the readily available information. If the fit exists, a personal interview will be arranged
to confirm the evaluation from the data.

Too few employment candidates make as much of an effort to
evaluate a prospective employer as it does of its candidates. Any
serious effort on the part of the prospective employee should include some research on the company which would be available in any
library. Its past financial performance, patents held, position in its
markets, the names of its executives and their backgrounds are
easily obtained. What its executives say and outside evaluation of
the company and its leadership are often available through the public
press. If this does not suffice, well-considered questions at the
time of interview should provide any missing information. How the
interviewer responds to the prospective employee's questions will also
reveal valuable information with which to assess the company and,
conversely, the candidate. The initial position and salary will soon
change, but one's personality and the culture of the company will be
slow to change, if at all.

If fit is important to a company, one will find reinforcement
or confirmation built into its development and evaluation system.
Periodic appraisals and evaluations provide the company opportunity

	Company description of me	Description difference
a. Risk Taking	_____	_____
b. Collaborative	_____	_____
c. Hierarchical	_____	_____
d. Procedural	_____	_____
e. Relationships-Oriented	_____	_____
f. Results-Oriented	_____	_____
g. Creative	_____	_____
h. Encouraging	_____	_____
i. Sociable	_____	_____
j. Structured	_____	_____
k. Pressurized	_____	_____
l. Ordered	_____	_____
m. Stimulating	_____	_____
n. Regulated	_____	_____
o. Personal Freedom	_____	_____
p. Equitable	_____	_____
q. Safe	_____	_____
r. Challenging	_____	_____
s. Enterprising	_____	_____
t. Established, Solid	_____	_____
u. Cautious	_____	_____
v. Trusting	_____	_____
w. Driving	_____	_____
x. Power-Oriented	_____	_____
Total difference	_____	_____

VALUE	WEIGHT	RATING		
		IND.	SUPV.	CO.

Theoretical

1	2	3	4	5

   x   ^   *

| | | 2 | 6 | 8 | 4 |

Economic

1	2	3	4	5

       ^   x*

| | | 2 | 6 | 8 | 8 |

Social

1	2	3	4	5

     ^x   *

| | | 2 | 6 | 8 | 6 |

Human

1	2	3	4	5

   ^   x*

| | | 5 | 10 | 20 | 20 |

Political

1	2	3	4	5

   x   ^   *

| | | 5 | 20 | 25 | 10 |

Moral

1	2	3	4	5

     ^x*

| | | 2 | 8 | 8 | 8 |

Individual ^   Supervisor *   Company x

Fig. 1.3 The value assessment.

to make these interim judgments. The employee has a similar opportunity to find out where he stands. He can determine his
strengths and weaknesses to allow reinforcement or correction. In
addition to the standard fare of appraisal formats, I suggest an
assessment of basic values. It would provide a means for the individual to judge where he fits with respect to company values.
The company can find out if its assessment of itself matches that of
its employees. Such an assessment is of critical importance to the
job seeker in his or her initial work experience. Socialization into
the company one chooses will certainly be much smoother if a high
degree of compatibility exists.*

Applying the assessment requires an examination of the similarity or differences that occur for each characteristic. Since some
are complementary and others direct opposites, summing frequencies
could lead to erroneous results. A sample table, as shown in Figure 1.3, might provide an initial judgment as to fit.

The maximum difference would be 72 and the minimum 0. On a
continuum, 0 would indicate the highest compatibility and 72 the
highest incompatibility. The assessment can be even more personal-
ized by weighting the various characteristics as to their importance.

The assessment system can be employed by an individual to determine where he fits within his perception of himself and the company. It can also be employed by management to assess how it
perceives itself and, through random survey, how it is perceived
by its employees. A better knowledge of the characteristics of an
organization's culture will enable you to better understand what
suggestions and ideas are acceptable and what techniques might be
successful in promoting them. It will also reveal areas of incompatibility upon which a decision must be made regarding change.

1.5   SUMMARY

Corporate culture is a composite reflection of its value set. This is
established by the company founders and modified or reinforced by
the corporate leaders who follow. What a company does and how it
is done are the manifestations of the value set. The levels reached
by an individual and accomplishments within that culture will depend
on how well the value sets are matched or accommodated.

---

*An excellent source for self-testing is included in the "Study of
Values" by Vernon, Allport, and Lindzey, published by the
Riverside Publishing Co., Chicago, IL.

Change is the most certain of all activities in a business en-vironment. Companies can initiate change or react to it. Which-ever it chooses, it means the adoption of new ideas and innovation. Whether those ideas and innovations are yours or those of someone else depends on two factors: the quality of the idea or innovation and how well it is sold. The latter is probably the most critical and it will be determined by your ability to judge the culture and how you you adjust and fit into it.

You will be either the initiator or responder to change. If you choose to be the initiator, the degree to which the change will be adopted is dependent on your ability to diagnose the receptivity to change and how it can be carried out. If your value set matches or is symbiotic with that of the organization, the task becomes easier. Whichever role you choose to play and how you play it de-pends on your own ambition and action and that of the company that will undergo it.

The preceding was meant as a means of gauging the responsive-ness to change. The management of projects is the management of change. Change is a threat to the status quo of most organizations. The importance of understanding the environment for change cannot be overemphasized. The remainder of the text is intended as a guide to how to execute successful projects. How they may be adapted to your own situation and how they may be implemented will depend on your ability to manage change within the culture you find yourself.

REFERENCES

1. Neil W. Chamberlain, *Remaking American Values*, Basic Books, New York, 1977, p. 20.
2. Jeremy Main, Waking up AT&T: There's life after culture shock, *Fortune*, pp. 66−70 (Dec. 24, 1984).
3. Thomas J. Peters and Robert H. Waterman, *In Search of Excel-lence*, Harper & Row, New York, 1982.
4. Ellen J. Wallach, Individuals and organizations: The cultural match, *Training and Development J.*, pp. 29−36 (February 1983).
5. Bernard Taylor and Gordon L. Lippitt (eds.), *Management De-velopment and Training Handbook*, McGraw-Hill, London, 1975.
6. E. Spranger, *Study of Values, Vernon, Allport, and Lindzey, a test published by Houghton Mifflin* (Translation by P. J. W. Pigors). Types of Men, Halle, Niemeyer, Ga.
7. Frederick G. Harmon and Garry Jacobs, Company personality, *Managment Rev.*, pp. 36−40 (October 1985).

8. Richard M. Hodgetts, in *Modern Human Relations at Work*, Dreyden Press, New York, 1984, pp. 265–268.

9. Robert M. Hodgetts, in *Modern Human Relations at Work*, Dreyden Press, New York, 1984, p. 269.

10. M. C. Grool, J. Visser, and W. J. Vriethoff, eds., *Project Management in Progress: Tools and Strategies for the 90's*, North Holland, Amsterdam, 1986, p. 303.

11. McKinsey & Co. learns some lessons of its own, *Business Week* (June 23, 1986).

12. Robert R. Blake and Jane Srygley Mouton, *The Managerial Grid III: The Key to Leadership Excellence*, Gulf Publishing Company, Houston, 1985, p. 12.

13. B. Taylor and G. L. Lippitt, eds., *Management Development Handbook*, McGraw-Hill, London, 1975.

14. Robert J. Graham, *Project Management: Combining Technical and Behavioral Approaches for Effective Implementation*, Van Nostrand Reinhold, New York, 1985, p. 4.

2

Planning

A plan is more than an objective and a means by which to achieve
it. It is a compendium of the path trod by its developer including
the objectives considered and rejected, the methods tried and failed.
It is a recognition that change may render the past unrecognizable
and the future uncertain. My admonition to students of project management is that in executing a project, they must leave a trail for
others to follow. In accepting a project in midstream, their first
obligation is to review whatever trail they find and assure themselves that the goals are achievable.

A project plan is the distillation of many facts and assumptions.
It is the chosen course of many considered and rejected. If these
are not preserved in some form, these paths may again be trod.
Often, rejected alternatives become viable as change alters the conditions for assessment.

Nothing builds more confidence in a project manager than the
knowledge of management that various alternatives, courses of action, and suggestions have been considered and weighed in the development of a plan. The complexity of both the internal and external environment demands an unfettered approach to the development of a plan. Nuclear power, hazardous waste disposal, and the
environment clearly demonstrate that there are constituencies which
cannot be overlooked in project planning. There is, however, a
time to fish or cut bait, when a manager must decide on a course of
action. Nothing can be gained without some acceptance of risk, but

that risk must be measured, its consequences predicted, and the plan designed with a resilience in response to inevitable change.

2.0   THE PLANNING ENVIRONMENT

> Management has no choice but to anticipate the future, to attempt to mold it, and to balance short-range and long range goals. [1, p. 121]

> Peter F. Drucker

The pursuit of management is profits. The quest becomes purposeful when encapsulated in a plan, which is contrived by its perpetrators, cycled to refinement under the watchful guise of its ultimate judges, yet adaptable to inevitable change discovered by its disciplined feedback.

2.0.1   The Purpose of Planning

No other mechanism communicates as effectively and completely as the statement of corporate purpose. Dupont's "Better things, for better living, through chemistry" and General Electric's "Progress is our most important product" are representative of the stated purpose of many of the world's corporations. Such statements tell the public the basic reasons for a corporation's existence. Within the corporation, purpose provides a means to marshal human and financial resources to achieve its objectives. It provides management a rallying force to focus the energy and capacity of its workers. Peters and Waterman, in their study of excellent corporations, concluded that one of the reasons for their excellence was their ability to extract extraordinary contributions from very large numbers of people through their ability to create a sense of highly valued purpose [2, p. 48].

Sense of purpose must be complimented by a sense of direction. Although the results may be subject to the vagaries of the uncontrollable, there is still much to be said for the attempt to influence the outcome by a considered and systematic analysis of the factors one can control. Human and financial resources may be wasted if the road to their pursuit is not adequately mapped.

Management provides direction in the form of specific goals for the long and short range. Most important, management must also reinforce this sense of direction by assuring that consistency is present in the form of deeds to support the words. Planning formalizes these objectives and consolidates goals, objectives, and resources in a time-constrained framework.

The plan incorporates a schedule wherein the achievement of objectives is fixed. The time constraint is essential as it focuses effort and aids in the determination and deployment of resources. It also forms the basis for adaptation should the plan require alteration.

2.0.2   Mobilizing the Resources

The genius of the industrial system lies in its organized use of capital and technology. [3]

John K. Galbraith

The successful organization is the culmination of planning by corporations that not only preach, but also practice, the planning process. They create an environment where ideas are nourished and culminate in actions toward known goals. People will enthusiastically engage in setting goals and priorities and work to influence the events of the future if they know they are a part of a larger effort in which all levels have contributed.

The uncertainty of the future and the certainty of change make any plan one of constant modification. As such, only the broadest goals seem to have continuity and all else is in a continual state of flux and alteration. People sense the need to remain alert to opportunities and thrive on the challenge of meeting new and ever-changing conditions. If a plan is beautifully printed and replete with elaborate charts and graphs, one may be sure that it does not represent what is going on or, what is often the case, nothing is going on [4].

There are two schools of thought regarding the planning process. The first is the bottom-up school, which believes management should establish broad goals and the ultimate plan should be a summation of the individual plans of each unit, department, and division. The second is the top-down school, which considers that plans should be established by management. The latter concept can be anything from the extreme of minutely detailed targets for production, profits, and costs, to general guidelines within which the lower levels of the organization are free to operate. Both approaches have advantages and disadvantages.

Proponents of the top-down methodology claim a shorter planning time cycle and a clearer communication of objectives. The bottoms-up proponents claim participation in setting one's own goals and objectives is what makes planning really work. This is one area where compromise can yield the best of both worlds. Since effective planning itself is a continual process, timetables to complete various aspects of a plan can detract from the overall objective. If deadlines are eliminated, more effort can be concentrated on what the plan is to achieve and how.

2.0.3   Focusing the Participants

It is better to plan for oneself, no matter how badly, than to be planned for by others, no matter how well. [5, p. 66]

Russell L. Ackoff

The benefit of having established one's own course cannot be overrated. Balance and direction can be achieved by management in the setting of targets and defining limits while yielding the decision on how best to achieve them to the individual participants. If a company is to unlock and benefit from the creativity of its employees, what better way than in the critical process of determining how it will fulfill its goals? An individual is most motivated when he can relate to the goals established and identify with the mechanics of how they are to be achieved. As with the participants in a tug-of-war, the objective is clear and the individual's contribution as part of the team effort is determined by himself, but discernible by all.

Project planning must take place in an environment subordinate, but complementary, to overall corporate planning. Individual projects are tangible objectives within the corporate plan. They may be physical, such as a capital expansion, rehabilitation, or renovation. They may be functional, such as efficiency improvements or procedural changes. This clear association with company goals enables project management to provide a balance of individual loyalties.

The very nature of a project provides its participants and management with tangible and clearly defined objectives. To build something within a specified time period and at a predetermined cost presents a far clearer target than, for example, seeking new markets for the company's products. The risk, of course, is that personnel associated with the projects' objectives will lose sight of the fact that seeking new markets is also an objective and may indeed be even more important. Project management must strike a balance to focus individual loyalties toward the immediate project objective within corporate objectives. Corporate management must, on the other hand, foster the benefits of project identification and encourage individuals to recognize how their immediate goals fit into the larger corporate picture.

Planning

2.1   THE STRATEGIC PLANNING PROCESS

2.1.1   The Objective

Ninety-nine percent of all surprises in business are negative.   [6, p. 94]

Harold Geneen

more a couple of options, No? question: fall back pl. The strategic planning process attempts to predict the outcome of various courses of action in achieving corporate goals.   The objective of this process is to assess, by simulation and analysis, methodologies available to a corporation in order to select those most likely to succeed.   A corollary to this objective is assessment of the company's environments, to judge reactions to its courses of action and to predict their influence.   Projects provide an area of high potential for alternative approaches and are therefore appropriate candidates for the strategic planning process.

Strategic planning addresses three questions:

1.   Where are we?
2.   Where are we going?
3.   Where do we want to be?

The answers to these questions are at the core of a business and require a formalized process and the highest-level involvement.

Planning cannot exist in a vacuum.   Boundaries must be identi-fied within which the planning must be confined.   Management estab-lishes these boundaries.   They are the direction in which the com-pany goals will be achieved and the limits of the resources that can be expended in the effort.   This direction provides the answer to two of our questions, but still leaves unanswered how these results are to be achieved.   Therein lies the dichotomy of top-down and bottom-up planners.

The Holistic Principle

This principle has two parts: the principle of coordina-tion and the principle of integration.

The principle of coordination states that no part of an organization can be planned for effectively if it is planned for independently of any other unit at the same level. Therefore, all units at the same level should be planned for simultaneously and interdependently.

The principle of integration states that planning done independently at any level of a system cannot be as effective as planning carried out interdependently at all levels.

This concept of all-over-at-once planning stands in opposition to sequential planning, either top down or bottom up. [5, pp. 71-73]

Russell L. Ackoff

The argument of who best to plan will no doubt continue. It would seem from the results, particularly those assessed by Peters and Waterman in *In Search of Excellence*, that the bottom-up school is currently most popular. The involvement of middle-level managers, those who make the machinery operate, is increasing [7]. That planning is necessary is no longer the issue.

Another issue within the planning concept is when to plan. Is planning a scheduled exercise which must be undertaken and completed in accordance with some preestablished timetable? Or is it a continuous process which has scheduled summary points to fulfill the formalized requirements of corporate administration? The former is supported by those who believe that the rigidity of imposed deadlines is necessary, not only to get people to plan, but to have results available for control. The latter has proponents who claim the creative and innovative processes cannot be turned on and off by the calendar. They claim a company cannot be constrained and must be responsive to challenge and opportunity.

History may repeat itself, but overreliance on prior action smothers the very creativity and innovation necessary to provide the competitive edge. It is necessary to know what is going on and what has or has not worked before. This reflection should stimulate, not inhibit, the imagination. It is by difference that measures are made, not similarity. The company that leads is most often one that balances the lessons of the past with the calculated risks of the future and takes the initiative.

The diversity of industries and companies within those industries makes it impractical, even presumptive, to suggest what, when, and how to develop a strategic plan. Such a plan is necessary, however, if there is to be a continuance of the enterprise and certainly if the company intends to grow within its industry. Planning involves the setting of goals and objectives, the establishment of a formalized process for its development and implementation, and the assignment of responsibility for these efforts. The development of a strategic plan will identify those projects which are necessary. It is through their execution that the goals and objectives of the plan will ultimately be achieved.

2.1.2   The Environment

Strategic planning . . . is the application of thought, analysis, imagination, and judgment. It is a responsibility, rather than a technique. [1, p. 123]

Peter F. Drucker

The ultimate course of action decided upon by any company should be the culmination of considered analysis of the alternatives available to it. Part of this analysis is consideration of the various environments in which the company must operate.

Competition — External organizations which offer goods or services of equivalent or alternate utility to the market.

The most apparent and obviously strongest influences on a company's operation by external forces are its competitors. In developing strategies it is necessary to know the bases from which the competition operates, what it is likely to do in the future, how and when and what resources it has at its disposal to affect such actions. Much of the information needed to make such assessments is readily available from catalogues, shareholder reports, the company's 10K report to the Internal Revenue Service, trade publications, and the public press.

Fortunately, most competition are members of trade associations and are publicly held companies. As such, they submit data and reports to these associations and regulatory bodies, which can be gleaned for useful information. From such data, production capacity, sales, employment, wages and salaries, and financial resources can be readily established or projected. Published photographs can be used to determine plant size and hence capacity. Purchase and disassembly of products can yield valuable information on production methods, raw materials, and quality. All of this can be legitimately used to improve one's own position or to emphasize strengths.

Political/Legal — Laws, rules, and regulations currently in force or likely to be enacted which impact the conduct of business.

To attempt to control or manipulate today's environmentally aware and sensitive citizenry, for example, is likely to turn a poorly informed project skeptic into an embittered project opponent. [8, pp. 37 – 38]

F. L. Harrison

The government has a significant impact on the conduct of any business. A large portion of a company's resources is directly involved in compliance with government regulations and reporting requirements. This effort adds to the total cost of production or services. It is not only necessary to consider, in its strategies, laws, rules, and regulations which already exist, but those which are likely to be promulgated. Most trade associations retain permanent lobbying forces both to collect information on pending litigation and to provide information to and influence legislators.

Sociocultural — Mores, customs, and attitudes of the general public.

The increasing need to give consideration to public opinion has not been lost on corporate managements. The acceleration of public awareness as a force to be considered by corporate planners began in the late 1950s and does not appear to have reached its peak. The reason for this phenomenon is beyond the scope of this text and is mentioned here only as an increasingly important factor to be considered in any corporate action. As an example of the impact of public opinion and action, one can cite the impacts on the nuclear power industry, the chemical industry and its relation to specific projects, and the delays and cost overruns of the Alaska pipeline.

2.1.3   Impact on Projects

A project, of whatever nature, must be considered as a means to increase profitability, market share, or other corporate objective. It may be conceived at any level in the organization, executed by it or others, and it may or may not be led or managed by its original proponent. As it was defined earlier, a project is an idea that has finite dimensions in scope, time, and money. To bring it to fruition it must have a champion and a place in the corporation's future. It therefore is an integral part of the strategic planning process. A project's champion or proponent must therefore determine the situational influences on the project, how they affect it, and what can be done to minimize the negative and maximize the positive consequences of the courses of action chosen.

The influences of competitive forces on a project are not only related to what competitors are doing, but are internal as well as external. A list of the important forces is given in Table 2.1.

Table 2.1   Internal and External Competitive Influences on Projects

Internal	Funding	External	Services
	Human resources		Labor
	Physical resources		Land
	Management attention		Funding
	Corporate philosophy		Materials

2.1.4   Internal Competitive Influences

Funding

The first hurdle any project faces after its technical or functional viability has been confirmed is the internal competition for funds. As is most often the case, there are more projects that appear to meet the company's financial return or other criteria than there are funds available. There must therefore be some criteria by which projects to be funded are selected. Whatever means is chosen, management must be certain that the cost criteria are accurate.

Projects fall into two major categories, mandatory and optional. Mandatory projects are those which must be undertaken to meet internally or externally imposed safety criteria or are required to maintain compliance with legislation. Optional projects are those undertaken to improve efficiency, productivity, or profitability or to retain or increase market share. Obviously, mandatory projects have first call on a company's funds, and what remains can be applied to the optional projects.

In many instances, projects begin as plans of corporate divisions to meet their objectives. They may be implemented by the divisions themselves or a central project organization. They will be assigned a divisional sponsor to manage the project or follow its execution and often be responsible for changes, including approving fund releases.

In most cases, projects are chosen under a system of capital rationing. That is, there are limited funds available and there must be some mechanism to choose which projects are to be funded from those put forward for consideration. Some form of profitability measure is usually used, but one must be acutely aware that there are also other, often informal methods for project choice.

The economic choice of funding projects usually hinges on whether the corporation can make more on its capital by investing it or executing the project. The decision on which projects qualify for consideration is determined by setting a rate of return on the project which must be equaled or exceeded. This is called the hurdle rate. Projects that fall below this rate may not be considered for funding. Under the conditions of capital rationing, projects are usually ranked by rate of return; the highest are obviously the best candidates for funding approval. Net present value (NPV) valuation techniques are most commonly used to determine the returns from various project alternatives. With the rules established, the competition begins and with it various hazards that must be addressed.

Creativity, if by that is meant undirected, unstructured, untutored, and uncontrolled guessing, is not likely to produce results. But a system which does not tap and put to use the knowledge, experience, resources, and imagination of the people who have to live with the system and make it work is as unlikely to be effective. [1, p. 271]

Peter F. Drucker

Nothing is more frustrating to a project manager or proponent than to see a perfectly viable project rejected, while a questionable one is funded. Equally frustrating to management are the projects brought forward whose only claim to fame is the ingenuity and creativity their proponents demonstrated in putting together their justification. To those who have not experienced the process from either perspective, welcome to the real world.

Aside from purely financial criteria for the making of investment decisions are the intangible values that are considered. While the benefits to be derived from such values are not always easily understood and certainly impossible to quantify, they are nevertheless real. For example, there are balances to be maintained between divisions or departments to maintain morale, there are key personnel to be retained, or there may be a location whose public might need to be stroked. It is important for management to communicate these aspects of valuation as part of the selection process and to be consistent in their application. Nothing destroys morale more quickly than to allow people to believe that their efforts have been subordinated to the caprice or arbitrary exercise of power by management.

The highest art of professional management requires the literal ability to "smell" a "real fact" from all the others and moreover to have the temerity, intellectual curiosity, guts, and/or plain impoliteness, if necessary, to be sure that what you do have is indeed what we will call an "unshakable fact." [6, p. 95]

Harold Geneen

It is understandable that project proponents will make every effort to see their projects to fruition. These efforts should emanate from an inherent desire to accurately and fairly present the facts in as unbiased a manner as possible. Management can foster such purpose by fair and reasonable evaluation, rational explanation for rejection, and when confronted with attempts at distortion, open exposure of such efforts and meting out appropriate discipline. All too often, management has created an environment wherein unnecessary effort must be expended to illuminate the true facts, or, failing the expenditure of this effort, marginal and questionable projects garner valuable funds.

Human Resources

The availability of funds is redundant without the human resources
necessary to properly expend them toward the objectives of the
project. Where these resources may come from and how they are to
be organized will be covered in the section on organization. The
purpose here is to emphasize the quality, quantity, and qualifications of the personnel required to execute a project.

> Research and development activities are those concerned
> with advancing the strategic state-of-the-art in functional
> areas and with developing a system of plans and products/
> services for the company's future. [9]
>
> David I. Clelland and William R. King

A project begins long before its form takes recognizable shape.
The genesis of a project has its seed in the purpose laid down by a
company's board of directors and its management's goals. Its
embryology begins in research laboratories, as an idea in a part of
the organization looking to improve its efficiency, a salesman in the
field looking to overcome a competitive advantage, or a simple opportunity for profit. For our purposes, a project begins when resources are required to convert an idea into reality and those resources must be managed.

The conversion of ideas into reality is the very essence of a
company's ability to survive and to grow. As such, the review,
evaluation, and implementation of ideas should involve the highest
levels of management. Therefore, our first human resource need is
that of management. The usual form of management involvement is a
committee for the periodic and systematic review of proposed projects. Because of the variety that projects can encompass, such
committees are made up of a cross-section from various functions in
order to bring to bear the best expertise and broadest viewpoints.
Its members should be the corporate management of the enterprise,
in order to emphasize the importance of such reviews to the future
of the company.

Part and parcel of the presentation of ideas for review and approval is the need to assess the human resource requirements the
project itself will require. This is particularly important if personnel are in addition to those which the project's proponent is prepared to commit to the effort. Strategic planning for these resources
should have included sources, numbers, qualifications, and contingencies in both cost and timing if these must be obtained by recruiting outside the company. The approval of a project is dependent not only on the validity of its objective, but on the quality
and completeness of its presentation.

The manning plan for the project should include a chart of re-quirements in calendarized form. Key individuals should be indicated along with their job function and the duration of assignment. Staff may be indicated by numbers of individual functional disciplines re-quired. If, through negotiation with other organizations, names of individuals can be supplied along with those the proponent is pre-pared to commit, these names should be part of the plan. This will ensure that assignment of those specific individuals gets the review and approval of management and its commitment to make those spe-cific individuals available for the duration of the project.

If specific individuals cannot be committed to the project, the requirements should include a list of the qualifications, experience, and level of expertise needed. This will enable functional review-ers of the plans to put forward candidates for consideration. It will also help to identify those positions which may have to be filled from external sources so as to obtain the necessary authority to recruit, should the project be approved. A typical manning plan is shown in Figure 2.1.

A project's success or failure will, for the most part, depend on the people assigned to it and their ability to work together. The project proponent would do well in developing its plan to include a maximum number of known key personnel and, in particular, a project manager with a proven record as a leader. This, of course, is not always easily achieved and some means must be available to assure success when a project must be staffed with a majority of unknown or untried personnel. This will be discussed in the chap-ters on organization and control.

Success in project management requires the wholehearted, sincere, constructive support of all the functional executives in the organization. Perfunctory announcements and direc-tives about a project will not achieve this. It takes direct personal signals from the top executive to the other top members of the team to convey the message that the project will succeed, and that all members of the team will be measured by its success. [10, pp. 31-32]

Charles Martin

What is found in attempts to staff projects is the reluctance of managers to release their best talent. Also, there is willingness to appear helpful by attempting to pawn off their poorest ones. This situation is prevalent in organizations that are not project oriented, or whose middle management is more concerned about its own sur-vival than in furthering corporate goals. Project proponents, in

POSITION	JOB DESCRIPTION	SALARY LEVEL	ASSIGNMENT DURATION
Proj. Mgr.	PM-2	E-6	1/90 -- 12/93

INCUMBENT (I) CANDIDATE (C)	DEPARTMENT AFFILIATION	NEW HIRE (Y/N)	COMMENTS
J.Jones (I)	Project Dept.	--	Retires 3/94

POSITION	JOB DESCRIPTION	SALARY LEVEL	ASSIGNMENT DURATION
Proj. Eng.	Eng.-4	E-3	2/90 -- 9/93

INCUMBENT (I) CANDIDATE (C)	DEPARTMENT AFFILIATION	NEW HIRE (Y/N)	COMMENTS
R. Smith (C)	Eng. Dept.	- -	Electrical Experience
B. Tuff (C)	Operations	- -	Currently Eng.-3

POSITION	JOB DESCRIPTION	SALARY LEVEL	ASSIGNMENT DURATION
Contracts Admin.	Eng.-3 or Fin.-4	E-2	3/90 -- 12/93

INCUMBENT (I) CANDIDATE (C)	DEPARTMENT AFFILIATION	NEW HIRE (Y/N)	COMMENTS
J. Roberts (C)	Purchasing	- -	
P. Graze (C)	Engineering	- -	not available until 4/90

POSITION	JOB DESCRIPTION	SALARY LEVEL	ASSIGNMENT DURATION
Controls Mgr.	Admin.-1	E-4	1/90 -- 12/93

INCUMBENT (I) CANDIDATE (D)	DEPARTMENT AFFILIATION	NEW HIRE (Y/N)	COMMENTS
E. Foss (I)	Engineering	- -	

Fig. 2.1   Typical project manning plan.

particular the project manager, are usually powerless to overcome this problem. Resolution therefore requires corporate management's interference and direction. As a starting point in its strategic planning, a project's proponent should lobby with its corporate leadership on the review panel for support and be prepared to name individuals it wants even though other organizations have been reluctant to release them. In the long range it is still better that corporate direction provide the incentive to middle management to cooperate in the manning of projects.

Physical Resources

Many projects, in their implementation, require the acquisition or reuse of land, physical plant, and machinery. These may be temporary or permanent needs. The project's strategic plan should include an assessment of these needs and a careful analysis of existing facilities in order to make maximum utilization of what may already be available for execution of the project. The schedule should delineate the acquisition and utilization of plant, what use will or can be made of such facilities when they are no longer needed, or when they will be returned to their previous use.

New plant acquisition will come under very close scrutiny since it usually represents high early spending, with high probability of nonrecovery should the project be aborted. The key planning effort should therefore be concentrated on maximizing use of existing plant, either through use of idle or spare capacity or modification at lower cost than acquisition. If this analysis indicates that the project's economics relies on new acquisition, then the recommendation should clearly outline the effort made to utilize existing plant.

Care should be exercised in planning around the utilization of existing facilities in that those who control them are aware of those plans. When the utilization has been negotiated, its timing and duration should be clearly indicated in the plan. In this way a contract is in essence created when the project is approved.

Management Attention

The reason behind the absence of focus on project or people in so many American companies, it would seem, is the simple presence of a focus on something else. That something else is over-reliance on analysis from corporate ivory towers and over-reliance on financial sleight of hand, the tools that would appear to eliminate risk but also, unfortunately, eliminate action. [2, p. 40]

Thomas J. Peters and Robert H. Waterman, Jr.

Management must have an active involvement in its projects. In some organizations, such as those in the engineering or construction industry, projects are a matter of survival. In others, where projects may involve only a small portion of the workforce, they may still be of significant importance as to warrant it.

It can be confirmed by many project managers, particularly those working in areas remote from headquarters and on physically impressive facilities, that they see more of the corporation's executives than personnel resident in the same buildings with them. Although visits by executives are useful and welcomed by the project's management, attention must go far deeper to have the desired effect, which is the knowledge by those responsible for the project that their efforts are appreciated, their needs are being satisfied, and help is available when needed, because management really cares about what is happening. It must be recognized that regardless of the business, they are all people businesses and it is people that make them go. People need to be motivated, given direction, allowed to reach beyond themselves, and occasionally, stroked or disciplined.

Part of the strategic planning for a project is how to navigate the charted and uncharted internal waterways. Equally important is knowing when to drop a loser and when to push a probable winner. This demands an early determination of costs, profitability, and ranking with respect to availability of funds and other projects. The more time and effort put into the development of a project, the greater the risk that justification will be biased to justify that investment. Management can assist greatly in this determination by frequent reviews of projects in a developmental stage. During this review, those requiring the knowledge should be made aware of the funds likely to be made available, the projects competing for funds, and their ranking.

Corporate Philosophy

How much attention and weight a project has will depend on the corporate philosophy toward projects. Corporations are either project driven or they are not. Aerospace, engineering, and construction organizations tend to be project driven. Their existence depends on the development and execution of projects and hence they influence the conduct of business. Those which are not project driven can run the gamut from those that place great importance on projects, such as a product developer, to those whose projects are few and far between.

When the fundamental objectives of a corporation are production oriented, the organization is geared to facilitate production. The introduction of a project initiates change which many will view as disruptive. If the corporate hierarchy does not fully support the project and communicate this support, the project will have difficulty in obtaining the necessary resources for its execution, even if it is adequately funded.

2.1.5   External Competitive Forces

The project manager who can execute his project with internal resources can find solace in the occasional difficulties in obtaining or controlling them when compared to the project manager who must forage for his needs in the outside world.

Services

Outside services can range in scope from contracting out the entire project to occasional individual consultations or experts. The breadth of services available boggles the imagination. The problem with services is not where to find them, but how to evaluate them and assess their cost. In developing the strategy for a project, the need for outside services must be included and their sources sought out, their capabilities and capacities assessed, and their costs determined.

Service companies and individual consultants either advertise their services or, in the case of those whose professional organizations restrain open solicitation, can be found through contact with those organizations. Enquiries will usually bring a deluge of information as to the scope of services and a listing of past clients. As a first screening, it is prudent to check with some of these clients as to their satisfaction with the services rendered. This can be most useful in pruning the list to those which should be further assessed by personal contact.

There is no substitute for assuring that contracted services are awarded to a firm or individual with the capability and capacity to carry them out. A thorough investigation is therefore necessary. These firms and individuals are soliciting business from many sources. They are never sure from which source the actual business may come. There are others in the marketplace for services and they too are also looking to obtain the best available.

Provision of services hinges primarily on the capabilities of individuals within the service organization. As part of an assessment and evaluation, the critical need is to solicit those key personnel who have the qualifications and experience sought and with whom you can work. The next step is to get the written commitment that these individuals will be assigned to your project, should it be awarded.

In many organizations, policy and procedure require open solicitation for goods and services from outside sources if the cost exceeds a certain amount. The reasons for this are many, not least of which are the limitation of opportunities for collusive behavior and conformity with legislation regarding trade, such as the Export Act for U.S. companies. There are also those who believe that the lowest costs are achieved only through competition. If these restrictions are imposed on the solicitation of particularly design and engineering services, the evaluation and selection of the competition becomes even more critical.

Many countries, particularly those in the early stages of development, require utilization of local enterprises. This can be limited to minor participation, but often prohibits use of nonregistered companies for engineering services. The objective, of course, is to develop or protect local enterprise, but this restrains choice and adds a further dimension to the selection problem.

Like many businesses, engineering, particularly of large capital facilities, is cyclical. Everyone seems to be in or out of the market at the same time. Such a situation does not often effect a go or no-go decision regarding the project itself, but does tend to magnify the problems associated with contractor selection. In a tight market, the contractors with the best talent tend to fill their capacity first. This results in having to choose a contractor of lesser quality or going with the better one who will increase capacity at the expense of diluting the very quality that caused you to choose him in the first place. The selection and evaluation process will be covered more fully in the chapter on organization and in the Appendix.

Labor

Planning requires consideration of two types of external labor, which includes manual, skilled, and professional personnel. These are labor that will be directly hired by the project and that which will be required by others contracted to perform services for the project. Both are important not only from a requirement standpoint, but in timing. Labor required by the project tends to be an immediate need and requires early preparation and prompt action. Labor required by others working for the project are usually further out in time and probably location and require predictions of conditions as they may prevail in the future.

The cyclical nature of projects often requires that staff be recruited from external sources. In preparation for this eventuality, project proponents should develop the requirements for recruiting, job descriptions, and experience levels of prospective candidates. Consultation with the personnel department should give an idea of

current lead times for various classifications of personnel, but if
this information is not readily available, it should be one of the
first items in the planning process. Recruiting a professional can
take anywhere from 3 to 6 months, from the time a position is ad-
vertised until it is filled. In some instances, when the position is
in a foreign location, this process may take a year or more. One
can see that, if a fully integrated and prepared project team is required at the time the project is approved, an early start is mandatory.

Those not normally in the process of recruiting should not over-
estimate the organization's ability to respond. Even in large organizations, professional recruiting is not a routine requirement,
and because salary levels are usually high and experience in the
current market low, there will be delays. In addition, most professional positions tend to have generated layers of rules, regulations,
and approval levels required for hire. Consultation with the personnel department by those responsible for project staffing should
be high on the early agenda.

Companies operating where legislation has created almost permanent employment for citizens are reluctant to add permanent staff.
As a consequence, the process to do so is intentionally cumbersome
and time consuming. Knowing this, project managers should think
twice about rejecting internal offers, even of personnel with known
shortcomings. At least in these instances, the weaknesses are
known and can be offset by the strengths of others. It is possible that, given a project environment and the support and counsel
of new supervision, poor performers can be motivated to become full
contributors.

Requirements for nonprofessionals or professional staff with low
experience requirements present a lesser problem in recruiting.
Unless it is absolutely necessary, from the standpoint of policy or
confidentiality, consideration should be given to obtaining such resources from temporary help organizations, or to assigning the work
to a service contractor.

Even though labor requiring outside recruiting may be done by
others, it is necessary to know the markets from which this labor
may be drawn. This is particularly important on large projects
where construction forces can reach large numbers and must be
drawn from broad geographic areas. An accurate estimate of costs
and project duration requires a knowledge of local labor rates,
union status, expiration dates of current agreements, the size of
the current workforce, and other projects planned in the same
time frame. The basis upon which conclusions and estimates are
made should be recorded so that they can be monitored for change.
This will be covered in greater detail in the section on controls.

An often overlooked aspect of labor is the need for training. This can result in considerable cost and create severe disruption in planned schedules. Training can include simple upgrading of skills to the complete development of fundamentals. Assessments of the labor market are essential in determining the extent of required training and its impact on project cost and schedule.

Land

Projects requiring land on which to build facilities face two major concerns: site selection and site acquisition. Despite the apparent sequence of the two, it is acquisition, or the reasonable possibility of it, which should be of first concern. A site can always be modified or its shortcomings offset. If it is unobtainable or is encumbered by zoning restrictions, its suitability is immaterial.

The purchase or leasing of land should be carried out with the utmost discretion. This can maintain its price and retain the ability to manage response of public reaction. Unless extremely large blocks are required, this is sometimes best handled through local agents who know the area and the zoning laws and who may be able to advise on the sensitivity of the public to the project. A lesson can be learned from the experience of General Motors in locating their Saturn plant. The relations campaign was such that several states were vigorously pursuing General Motors to build the plant in their states with offers of land, loans, and other inducements. This is not to say that localities will be competing to get you to build your hazardous chemical plant in their towns, but it does indicate that stressing the positive aspects of a new plant, such as employment and local expenditures, can be of great assistance in gaining acceptance, if not removing obstacles.

Alternate locations, ranked by preference complete with details on cost, zoning problems, timing of the acquisition, and probable public reaction, should be proposed. A plan to address the consequences and problems associated with a switch to an alternate should be included.

Material

Availability and cost are the key elements to be considered in the planning of the material requirements of a project. Shortages and delays in procuring material are but two factors affecting the plan. The cyclical nature of most commodity businesses exacerbates these problems. Costs, in particular those associated with building projects, can represent the major project cost element. The longer duration over which materials expenditures usually occur makes them the most difficult to predict.

Developing strategy for material procurement is best started by dividing requirements into two groups: items requiring long lead time and those normally obtainable on short notice or usually available from stock. Efforts to ensure availability when needed and at a reasonable cost can then be concentrated on the long-lead-time items.

Long-lead-time items fall into two categories: engineered items which have to be designed and developed specifically for the project and those which are in short supply or have a long manufacturing cycle. Highly engineered materials, including equipment, are usually the core of the project and are the key to its success and performance. They will therefore deserve the greatest attention. This is particularly true should the equipment or material be "state of the art," in other words, stretching current technology. The tendency is to be overly optimistic as regards both cost and the duration of the development and production cycle. This may be overcome by obtaining estimates from neutral experts, thorough review and assessment by management, and inclusion of additional contingency for both cost and time. It usually requires all three, and even this does not, by any means, ensure complete accuracy.

Manufacturing cycles are a major element in the material schedule over which there is but minimal control. Strategy, in this regard, should be aimed at minimizing the potentially adverse consequences of this lack of control. There are several ways that can be accomplished and these should be assessed within the schedule and cost aspects of the project.

Pilot Studies. In many cases, process facilities whose future viability is highly uncertain can be downsized for a more modest investment and utilized to develop more accurate capital, production, and operating cost factors. They are useful also to uncover potential problems. In some cases, existing facilities can be modified with the same result.

Operating Factors and Sparing. Equipment sizing is determined by the specific requirements of the project. The capacity required can then be met within parameters dictated by the operating factors* developed for the project. High operating factors are associated

---

*Operating factor is expressed as a percentage arrived at by dividing the time in days the plant can be expected to run at full capacity by 365 days per year. Operating factors normally range between 75% and 95%. They are determined by methods which utilize the experience in operation of various individual elements in a plant, such as pumps, motors, controls, etc., and are beyond the scope of this text.

with facilities whose output must be secure or where the economic return on the extra output far exceeds the higher cost. High operating factors for processes that utilize equipment at the limits of technology, are subject to frequent breakdowns, or have high levels of maintenance are obtained by sparing, installation of several units whose capacity adds to 100% of requirements, or a combination of the two.

When more than one unit to satisfy a process requirement is to be ordered, opportunities are available to overcome the potential problems associated with production delays. If the unit required is one that may be subject to many or significant changes as the project develops, it may be prudent to incorporate an imposed delay in the production cycle between units. This would enable early delivery of at least a part of the requirements, ensure that required changes are incorporated into the later units at probably lower cost, and minimize the number of modifications that may have to be made on site, at always higher cost. Another possibility is to spread the risk of delay by ordering units from separate suppliers.

Each of these factors requires separate analysis within the operational and cost criteria of the project. There are various ways to achieve desired operating factors. One is utilizing two half capacity or three one-third capacity units. Spare-parts policy and standardization also must be taken into account. Sensitivity analysis should be a mandatory requirement at this point regarding the level of operating factor chosen. Much of the capital investment is dictated by this factor and management should be intimately involved in its determination.

Supplier Experience. Nothing can replace the experience with suppliers as a determinant of the probability of on-time delivery and response to changes. Past performance is a reliable gauge of what can be expected in the future and even if poor, as regards delivery, it enables a judgment as to what is probable. Relationships with suppliers also affect reactions in response to delays or technical problems.

The experience gained in dealing with various suppliers is similar to that gained in working with one's peers or subordinates. Their strengths and weaknesses become known and can be accommodated. Given this, it is well to consider repeat business, particularly with those orders requiring a high level of cooperative effort. On the supplier's side, it is safe to say cooperation is usually better when the prospect of future business is in the balance.

Funding

Project financing can be thought of in a systems sense and placed within an overall project-planning project-

management function, but it cannot be thought about outside of the context of the real world of money. [11]

Albert J. Kelly

Project managers, except for those managing joint venture efforts, are not usually involved directly in obtaining funds for their
projects from external sources. This is a function of the finance
department, and the methods and sources of external financing are
outside the scope of this text. That a project manager has no responsibility in outside funding operations is, however, a fallacious
assumption. The project manager has a responsibility to determine
the amount of funds needed and the timing of expenditures. The
finance organization, in turn, has the responsibility to obtain and
dispense funds and to communicate to the project the cost of these
funds and the opportunities open for alternatives in financing the
project.

Whether borrowed from outside sources or supplied from internal cash flow, the cost of these funds or the opportunities available from other uses are important factors in the management of
projects. Irrespective of interest rate levels, vigilance is required
to ensure that only those funds actually needed are requested and
expenditures deferred until absolutely necessary. In times of widely fluctuating interest rates, this may need to be modified, but it
is a good general rule. The cost of funds is not usually charged
directly to the benefiting project, but is assessed as part of general
corporate overhead. This does not negate the responsibility of
project management to keep these costs to a minimum.

The intimate relationship between the management of the project's
expenditures and the management of its financing requires greater-
than-usual cooperation between the project and finance organizations.
This is indeed unfortunate, since the benefits of courses of action
taken by one can be entirely offset by the actions of the other.
For example, the company may be accumulating currency of one
country through sale of output, while the project is buying commodities from the same country. Knowledge of these activities may
enable the hedging of risks of currency fluctuations. Opportunities
can also be lost because of the failure to communicate. Companies
that operate in the international market would especially benefit
from increased cooperation. In some instances, local currencies can
be used for procurement which are otherwise blocked from repatria-
tion. Where large purchases in foreign currencies are to be paid in
the future, currency risks may be minimized by forward buying of
the amount needed, options, or hedging the risk.

Opportunities are available to reduce overall project funds by
taking advantage of special financing arrangements, tax benefits,
and worldwide procurement. Countries, states, and municipalities

have developed innovative financing schemes to attract job-creating projects. They include: complete low-interest financing of the total project, partial financing, interest subsidies, and the financing of material and equipment purchases. Tax holidays may be offered for locating a project within the area of the taxing authority. Opening project purchasing to worldwide supply can result in lower prices. Many products are manufactured to uniform or nearly equivalent standards, and a minimum sacrifice in reducing standards or specifications can yield significant savings. A drawback of this process is the added time needed to develop these sources.

The planning for the project should, at a minimum, include a schedule of when expenditures will be committed and when they will be paid. As these occur, these schedules should be regularly and routinely updated. The timing intervals should be determined jointly by the project and finance organizations. More detail on this subject will be provided in the section on controls, but it is important to recognize here that there is difference between the funds predicted to be spent and those which will actually have been expended by the end of the project. This difference represents the estimating error and the degree to which contingent funds are expended. The management of these differences and its impact on project costs is an important element of cost control.

2.2   BUSINESS PLANNING

2.2.1   The Long-Range Plan

Planning is a function of all managers, although the character and breadth of planning will vary with their authority and with the nature of policies and plans outlined by their superiors. It is virtually impossible so to circumscribe their area of choice that they can exercise no discretion and unless they have some planning responsibility, it is doubtful that they are truly managers. [12]

Koontz, O'Donnell, and Weihrich

The Planning Horizon

Goals and objectives must have some definitive time frame in which to be reached or achieved. Without such a time frame, direction is clouded and actions flounder in search of the finiteness that gives purpose to control. Those which are in the near term require careful definition and reasonably rigid parameters. Those in the longer term cannot be so defined and must be flexible to allow adjustment to changing conditions. Reexamination on a regular basis is essential to monitor applicability to current reality and to maintain vitality as a guide to action.

Planning periods, or horizons, range from 6 months to 3, 5, and even 20 years. In one case, a Japanese firm was said to have a 100-year plan. The further out in time, the more difficult it is to forecast conditions that may prevail, where the organization may be in relation to its markets, and its relevance to the economy as a whole. Peter Drucker has said, "Planning is necessary precisely because we cannot forecast" [1, p. 124]. As you may have discovered in attempting to use a 5-year-old road map to plan a trip, change is certain, but it hasn't prevented the publication of road maps.

The company plan is its road map. The time frame used most often depends on the volatility of its industry and the magnitude and variety of changes that can be expected. If product turnover is rapid, adjustments to market conditions frequent, and response times short, the planning horizon will also be short. For example, the garment industry has four annual seasonal changes to which it must respond. Projections of what the market may want and the ability to respond with fabric design, garment design, manufacture, and distribution are subject to the unpredictability of style and fashion. The planning horizon is therefore short and may be anywhere from 6 to 12 months. The forest products industry, on the other hand, is working with raw material that may take anywhere from 30 to 50 years before it is ready for market. Its long-term planning must be adapted to this constraint.

The most common planning horizon for manufacturing, utility, and service industries is the 5-year plan. The 5-year goals and objectives are loosely defined and adjusted only if conditions warrant. The near-term or 3-year objectives are more definitive. Projects in this plan are likely to be implemented in the next year or so, modified as necessary, or dropped. The first year of the plan is usually called the operating plan and forms the basis for actions to be undertaken, continued, or discontinued as soon as the plan is approved by management.

> Long range planning should prevent managers from uncritically extending present trends into the future, from assuming that today's products, services, markets, and technologies of tomorrow, and, above all, from dedicating their resources and energies to the defense of yesterday.
> [1, p. 122]
>
> Peter F. Drucker

The planning of events based on repeatability of historical experience has gone unchanged for many years in most companies. Its applicability under today's conditions is open to question and,

if adjustments have not been made, could lead to planning for failure as opposed to success.

History, as a primary basis for assumption of future results,
can be a trap rather than a guide. The project that took 3 years
to accomplish 10 years ago, now takes 5 years. Indeed, the project
just completed may now take half again as long if duplicated starting today. Increases in cost may go far beyond that explainable by
inflation. The primary causes of this extension in project duration
and cost are: compliance with ever-increasing regulation, litigation
or other forms of involvement by the public and environmental
groups, and labor productivity. Planning must take these factors
into consideration and one way to accomplish this is to include contingencies in cost estimates and time contingencies into duration
estimates.

Objectives/Goals

> Good project management focuses on goals; other consid-
> erations are secondary. Such a single-minded concentration
> of resources greatly enhances project success. [10, p. 27]
>
> Charles Martin

A plan must have focus and direction. These elements enable
resources to be applied efficiently and purposefully. To provide
them, management sets goals to work toward and objectives to be
achieved. Planners then develop complimentary subsets which guide
the development of specific plans for each segment of the organization.

Goals reflect the purpose of an organization's existence. They
focus the efforts of the organization and function as a measure of
its accomplishments. A specific goal should be such that it may
never be satisfied, while at the same time remaining a target for
achievement or maintenance. Market leadership is such a goal. One
can attain a higher percentage of the market than any competitor,
but unless it is 100% and not growing, there is still more to be attained. Other similar goals may relate to product quality, customer
satisfaction, brand awareness, or business volume.

Objectives are specific ends that will partially satisfy the ultimate goal. They can be a percentage of market share, dollars of
sales, percentage return on sales or equity, or number of rejects on
final inspection. They are elements of the plan that can be classified by measurable performance within fixed time periods. They
are identified with expenditures that reflect the price the corporation is willing to pay in order to achieve them. Within this framework, individual organizations and individuals have complementary
supporting objectives.

Several characteristics identify well-conceived objectives:

1. There is broad participation in their establishment.
2. They can be achieved with reasonable effort.
3. They involve the cooperation of two or more departments of the organization.

Workers at all levels are more highly motivated when they have participated in the setting of their own objectives. Enlightened management fosters a climate where this participation can occur and encourages individuals to become involved. Maslow's hierarchy suggests that humans have several needs besides the basic needs of food, shelter, and security [13]. Two of these can be described as the need to belong and the need to stand out. The apparent contradiction is easy to understand when one reflects on the actual observation of many of the people we have met or worked with, including ourselves. Given a challenge we will work together toward the common objective. This is teamwork, and we all like being identified with a winning team. Ask us individually where we believe we fit in the group and the response will almost universally be above average. It confirms the need to stand out. One cannot be on the winning team or stand out if the standard is set so high it cannot be achieved, or the failure rate is high. Management can ensure winners by establishing achievable objectives. If it wants to enhance teamwork and cooperation, it will encourage objectives that require departments and divisions to work together in a cooperative effort. Some companies even set objectives lower in order to ensure that a majority enjoy the elation and self-satisfaction that come from meeting or exceeding them.

Figure 2.2 illustrates the formation of organizational and individual objectives around the core of corporate goals to form a cohesive whole.

Business Conditions

No long-range plan is complete, or even properly constructed, without reference to the business conditions on which it is based. The plan's goals and objectives must be founded in an assessment of how management views the larger environment in which the company operates. Such elements as the general economy, the political outlook, and economic and social trends all influence operations. Changes in these elements and their effects must be forecast.

Preliminary to the preparation of the long-range plan, management must give the planners its view of the business environment, conditions as they perceive them, and a projection of what these are likely to be in the future. This outlook should include not only economic factors, but any other that influences the company's

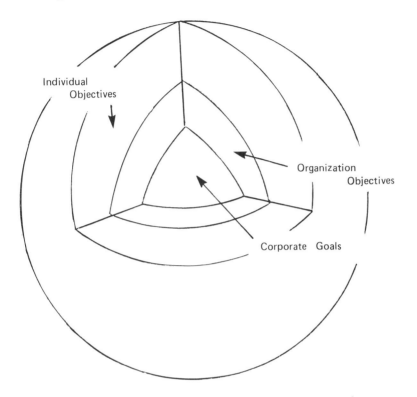

Fig. 2.2   Form of objectives and goal interdependencies.

future.   In turn, planners can refer to these considerations in the
establishment of objectives.

2.2.2   The Operating Plan

Because the only certain thing about the future is its
uncertainty, anticipating and managing change —the essence of planning —is a vital managerial function.   Indeed,
it is the most crucial of all functions.   [14]

M. K. Badawy

Establishing the Objectives

Projects form an integral part of company objectives.   They may be
carried out some time during the long-range planning period, or

they may already be in progress. As part of the overall plan, individual projects conform to the corporate goals and may result in the achievement of one or more of its objectives.

Regardless of its organizational form, many company objectives are carried out as projects. There may be a permanent project-type organization, or a project group formed on a part- or full-time basis to carry out a specific assignment. A project is characterized by its objectives and identified by a budget and time constraint. This identification may be vague, such as the time constraint in the basic research for a new drug or the development of a new product. Or, it may be firm, such as the building of a plant. Within these objectives, the personnel assigned to the project will have individual objectives as part of their responsibilities within the project team.

The very nature of a project and its clearly defined objectives obscures the need to remain cognizant of other corporate objectives. It is important that these be considered in the development of plans for individuals or the team. Professional development, paperwork reduction, resource efficiency, and technical development are continuing corporate objectives which can be adopted by individuals and the team.

Objectives, not to be confused with goals, should have three characteristics: clear definition, fixed period for accomplishment, and capability of feedback —in other words, what has been accomplished, when, and how it has been progressing during its implementation. If an objective does not meet these criteria, it is incapable of measurement and cannot be called an objective.

The operating plan for the project is a summary of individual and collective objectives. It includes the budget, timetable, and a means by which to judge progress. These complement the project and conform to corporate goals. It is a plan by which actions are initiated, results are compared, and personal performance is measured.

Scoping the Tasks

The need to make decisions on projects with a minimum of definition presents a constant challenge and increasing risk. It places a premium on experience in defining a project and estimating its cost and duration. Assumption must substitute for definition and provide useful reference.

Identifying the objectives generates the needs and opportunities inherent in developing how those objectives will be achieved. In the five-year plan, these opportunities are explored and developed. In the operating plan, they are refined and implemented.

A first run at planning project execution should be made without financial, personnel, or physical resource constraints. The first step is dividing the project into individual or group tasks.

Three benefits result from this approach: a check on the given budget and schedule, identification of opportunities, and uncovering of problems. The resultant plan then stands a greater chance of success, staying within given constraints. This prevents plans from going ahead which cannot pass such critical review.

The tasks should be such that they can be assigned to individuals or a group within the team. This provides an objective for performance measure and a motivating force for individual effort and group cooperation. It also enhances project control. Participation of the team members in this effort provides a further means of increasing motivation.

Assigning Responsibility

> No one follows the intent of plans as carefully as their authors, and no one resents plans as much as those who inherit them. [15]
>
> Robert D. Gilbreath

Dividing the project into tasks is the first step in assigning responsibility. The second is identifying those tasks with individuals or groups. Next to the selection of individuals to staff the project, this is one of the most important functions of the project manager. Success or failure of the project rests on the performance of its people. The performance of the people depends on their ability to meet the demands of their assigned responsibilities. This, in turn, is a function of capability and motivation. It is up to the project manager to match the tasks with the personnel. Capability relates to one's intelligence, skills, and experience. It is a time-stable element in that it undergoes change at a relatively slow, steady rate. Motivation, which often strongly influences our perception of capability, can undergo wide variations in rather short time spans. The objective is to maintain it in a positive direction and at a high level. Assigning the right tasks, at the right time, will help ensure this result. This can be achieved by keeping in mind the following points:

1. Control over one's own destiny. Giving the individual an opportunity to participate in the selection of assignments.
2. Growth and development. The task should take advantage of existing skills, experience, and knowledge, but provide room to stretch and attain new skills and knowledge, with an opportunity to advance one's position. The individual should be encouraged to seek assistance on his own, but supported to ensure that the risks of venturing are minimized.

3. Achievement. The task should be such that there is a high probability it will be successfully undertaken. It is equally important that when this occurs, it is duly recognized in some tangible way.

4. Social orientation. It is important to know whether the individual works better on his own or within a group.

The importance of matching tasks to individuals cannot be over-stressed as it is most often the reason for project failure. This will be covered in greater detail in the chapter on managing the failing project.

Setting the Timetable

At the time of establishing the project in the long-range plan, a duration was envisioned for its execution. When it becomes part of an operating plan, confirmation of the original time estimate is essential. This can be accomplished, as with the budget, when the individual tasks required to perform the project have been identified and their interrelationships determined.

Each task has its own duration. It may be executed independently from other tasks or be dependent on the completion or parallel execution of other tasks. Its duration may be accurately predicted or subject to considerable uncertainty. There are methods to handle either situation, but our concern at this point is that a task time frame can be determined and interrelationships shown such that an overall schedule of the project can be developed. As with the budget, this will confirm the original schedule premise, uncover problems, or provide opportunities.

Establishing Controls

Unfortunately, the development of sophisticated techniques that have made it possible to effectively plan and control large and complex projects, has also led to a concentration on these techniques and an inadequate recognition of the human factor involved. As a result, project planning systems often fail, many people have become disillusioned with formal planning and a great many firms do not use planning as effectively as they should [8, pp. 32–33]

F. L. Harrison

Not surprisingly, the majority of project controls are identified when the budget and schedule have been developed. There are other controls of equal importance, and care should be exercised as to the use of controls, the numbers monitored, and by whom. The great tendency is to overcontrol when the project controls themselves should be established in such a way to be self-controlling.

The main controls of projects relate to cost and schedule. These details are the most visible and the easiest to develop and monitor. They have a tendency to overshadow other necessary controls which are equally important, such as design, drawing production, award of contracts, procurement, and construction progress. In most cases, these activities are independent of forces outside the project and are a good yardstick of the performance of the project's management. Many projects cannot enjoy the luxury of having technical specialists assigned on a full-time basis. Periodic review at preestablished milestones provides the necessary control to ensure technical viability of the project and evaluation of critical specialized equipment or processes. Control of plans for personnel ensures that they are being implemented and enhances confidence in the employees that their efforts are being recognized.

Controls are best established at the level where action can be taken promptly and with minimum effort. The plan mandates the expected level of performance, for example the budget. A range about this level should be established, within which the project's management can control its activities, without taking away its freedom of action. This might be ±5%. Each subsequent level of management should have its own range, such that each level in the hierarchy becomes actively involved in decision making when the project reaches its level of control.

Feedback

> The usual situation is that there are are many versions
> of reality and data provided reflect an approximation of one
> version of this reality — maybe.   [16]
>
> Robert J. Graham

A system cannot control without feedback. Feedback provides the necessary communication of the results of action in order that status can be determined and reaction considered. Such reaction can be active or passive based on the preestablished range identified for control. A simple analogy illustrates the concept.

In order to maintain a comfortable temperature in a home a thermostat is used to control the operation of the heating and air-conditioning system. The system has a built-in bias limited by a maximum and minimum temperature setting, usually a few degrees apart. If the temperature exceeds the maximum setting, the air conditioner operates. If it drops below the minimum, the furnace comes on. The apparatus is thus self-controlling within preset and acceptable limits. The only time we, management, become involved

is when the system is not properly responding to the established
control. Then, we either make our own diagnosis of the problem
and get out the toolbox or call in expert assistance. As with
projects, one often leads to the other when we find ourselves unable
to solve the problem.

Management by exception is a method of control espoused by
progressive managers. All too often it is paid only lip service.
Nowhere is this more evident than in the area of feedback. Reams
and reams of paper are being generated in the name of control,
when in fact little gets read and the important cannot be gleaned
from the unimportant. The ease with which this can be accomplished has compounded the problem of paper excess. Worse, it
has provided a crutch for management to remain desk bound, when
they should be out in the trenches, finding out what's going on.

Feedback must be selective, pertinent, and timely. Part of the
plan should include a description of what information is to be provided for this purpose, who is to receive it, and when. It is also
important to establish what is to be done about it in reference to
who has control. In this manner, each level can function as intended and those required to know can be kept informed.

2.2.3   The Contracting Plan

Very few organizations can execute a project without some level of
outside assistance. Engaging outside assistance will involve contracts. The number of contracts on major projects can run into the
hundreds and may involve millions of dollars. Any project that requires contracting for goods or services should have a planned approach to what is to be contracted and how it is to be contracted.
A plan affords an opportunity to dissect the project into appropriate contract parts and to assess the method of contracting.

A contracting plan provides the answers to the following
questions:

1. What is to be contracted?
2. Why is it being contracted as opposed to being done "in-
   house"?
3. How is it to be contracted? Negotiation? Competitive or
   selective bidding?
4. With whom is it to be contracted, or how are the contractors
   to be selected?
5. When is it to be contracted?
6. How is it to be priced? Lump sum? Unit price? Cost reimbursable? Or some combination?

The plan gives management an opportunity to review the project and provide its input or acceptance at an appropriate stage in the project's execution.

Planning should include an eye toward the contracting requirement. Consideration must be given to where the work is to be performed, the contractor market in numbers and size, and other factors that may govern contracting strategy. The objective is to assure on-time completion at the lowest cost, consistent with the quality and performance requirements of the project. A work breakdown structure will be developed such that its elements can be isolated or combined to provide logical contracting packages. These packages will reflect the earlier considerations. For example, if the electrical portion of a project is estimated at $2,000,000 and the maximum annual volume of any individual electrical contractor in the project area is $2,000,000, several alternatives may need to be considered to avoid overtaxing a single contractor's resources. It may be prudent to consider expanding the contractor base to include those operating in a wider geographical area. If one needs to consider local public relations, it may be more appropriate to break the job down into two or more contracts within the normal capacity of local contractors.

It may seem obvious why some work is contracted and some is done with a company's own resources. This decision is not always easy, particularly when a company has a large maintenance work-force. The existence of such a workforce presents its management with the argument of capability, if not often capacity. On the other side, the project's management and perhaps the operating group who will benefit from the project see a loss of control of the project's execution. This generates an internal conflict of who is to do the work.

There is no doubt that most internal maintenance-type organizations have slack in their organizations. This is expected, inasmuch as such organizations must be able to cope with both routine as well as emergency or special activities. Maintenance management is, or should be, actively seeking productive outlets for the built-in idle time. Putting these resources to work on projects is a natural alternative. By the same token, project management fears the loss of control, in that these resources are under other direction and could be easily diverted if an emergency arose which has priority over the project. The objective and concerns are very real and must be solved by management. The compromise solution, when outside contracting is deemed inappropriate, is to ensure direct control by the project of the resources and, in an emergency, a fall-back plan which recognizes the disruption in both cost and time to the project.

Contracts may be negotiated directly or bid competitively in an open or selective market. Although the reasons for selecting an approach will vary with each organization, the primary reasons for utilizing a particular approach are given below:

Direct Negotiation — Individual experts with special skills for a particular job. Emergency work where time is of the essence.

Selective Bidding — Whenever it is desirable to restrict the contractor list to those most qualified in either or both aspects of experience and capability. A level of competition is introduced to attempt to secure a lower price than normally attainable by direct negotiation.

Open Bidding — All other situations. This may still be selective, in that bidding may be restricted to only those who are technically and financially able to perform the work, but open to all so qualified. Theoretically, this form of contracting obtains the lowest cost. It assures only that the initial price will be the lowest. The majority of U.S. federal, state, and local government contracts are secured by this method because of legal requirements to do so.

Within each of the above formats, the contract pricing methods can include lump sum, unit price, reimbursable cost, or any combination of the three. The benefits and shortcomings of these are discussed further in Chapters 4 and 5.

The open bidding format provides the least opportunity for improper or illegal conduct on the part of company employees and hence has associated with it the least controls. Few questions are raised when the contracting plan includes open bidding of all proposed contracts. As soon as selective bidding and in particular negotiated contracts are recommended, the proposer must be prepared to justify the selection. The possible consequences, be they potential for misconduct or complaints by the contractor community omitted from the selection, warrant the required justification.

Picking the contractors with whom the project will be executed is one of a project manager's most important tasks. Assuring that whoever is ultimately selected is fully qualified to carry out the task assigned can be a full-time job. In this effort, one reaps what one sows. As with any prejudgment, objectiveness is made difficult by the subjectiveness of the issues to be weighed.

What is best? Who can assure that a contractor, using the same resources, can repeat previous success? These are honest questions that have no easy answer. Decisions, however, must be made. There are methodologies that have been developed and refined over long usage for the purpose of contractor selection. They are not perfect and don't claim to be but they have a good record of re-peatable successes. One such method is given in the Appendix.

The contracting plan should include all of the above elements for all of the work to be contracted. Even if actual award of contracts requires further management approval, approval of the process that will lead to that decision will result in maximizing the positive effort and minimizing recycling. The importance of the contracting process in the execution of projects will be borne out in further sections covering staffing and controlling.

The contracting plan must include a schedule for each contract. This enables time and resources to accomplish the following steps:

1. Development of scope, specifications, and contract terms.
2. Preparation of a contractor slate, solicitation of interest, and selection of a bid test.
3. Solicitation of tenders, evaluation, and contract award.
4. Contractor mobilization.

2.3   SUMMARY

Planning is an essential corporate function. Effective planning can only occur if there is a fundamental purpose, participation at all levels is welcome and encouraged, and the results of the effort are used to guide the corporate destiny. Planning is a continuing effort requiring utilization of the best of a company's talent and focusing their energies toward realistic goals.

Strategic planning requires all the resources and ingenuity a company can bring to bear. Its effective application demands a clear understanding of corporate goals and knowledge of the environment in which it operates. In applying the process, consideration must be given to internal as well as external competitive forces. The rivalry for funds and human and physical resources is often overshadowed by the competition for management attention. External competition compounds the problems of selecting the proper strategy in pursuit of results. Hurdles, however, can often be converted into opportunity by skill and imagination. Diligence and perseverance in gathering and utilizing available intelligence are usually rewarded with success.

The long-range plan lays down the road map by which a company's goals and objectives are to be achieved. It reflects what is expected to be done, given the resources available and under the conditions the business anticipates it will operate.

The operating plan for a project presents the broad outline of its objectives, how they are to be achieved, and what resources are going to be applied toward their fulfillment. In the process of its development, original assumptions are tested, budgets confirmed, and the whole broken down into workable parts. These parts, in turn, are matched with the resources in a workable and manageable conglomerate. The plan includes financial, schedule, and operational controls, including feedback to augment its management at all levels.

The contracting plan is one of a project manager's most important activities. Selecting project elements to be contracted, developing contracting strategy and types, and choosing the contractors require careful consideration and can involve significant effort. The objective is to assure that the type of contract method chosen and the contractor selected will promote project success.

REFERENCES

1. Peter F. Drucker, *Management Tasks and Responsibilities*, Harper & Row, New York, 1974.
2. Thomas J. Peters and Robert H. Waterman, *In Search of Excellence*, Harper & Row, New York, 1982.
3. John K. Galbraith, *The New Industrial State*, Hamish Hamilton, London, 1968, p. 354.
4. Bertram M. Gross, *Organizations and Their Managing*, The Free Press, New York, 1964, p. 578.
5. Russell L. Ackoff, *Creating the Corporate Future*, John Wiley & Sons, New York, 1981.
6. Harold Geneen, with Alvin Moscow, *Managing*, Doubleday & Co., New York, 1984.
7. Louis A. Allen, Managerial planning: Back to basics, *Management Rev.*, pp. 15−20 (April 1981).
8. F. L. Harrison, *Advanced Project Management*, John Wiley & Sons, New York, 1981, pp. 37−38.
9. David I. Clelland and William R. King, *Systems Analysis and Project Management*, McGraw-Hill, New York, 1975, p. 196.
10. Charles Martin, *Project Management*, AMACOM, New York, 1976.
11. Albert J. Kelly, *New Dimensions of Project Management*, D. C. Heath & Co., Lexington, MA, p. 61.
12. Harold Koontz, Cyril O'Donnell, and Heinz Weihrich, *Management*, McGraw-Hill International, Tokyo, 1980, p. 159.
13. Abraham H. Maslow, *Motivation and Personality*, Harper & Row, New York, 1954.
14. M. K. Badawy, *Developing Managerial Skills in Engineers and Scientists*, Van Nostrand Reinhold Co., New York, 1982, p. 234.

15. Robert D. Gilbreath, *Winning at Project Management*, John
     Wiley & Sons, New York, 1986, p. 135.
16. Robert J. Graham, *Project Management: Combining Technical
     and Behavioral Approaches for Effective Implementation*,
     Van Nostrand Reinhold, New York, 1985, p. 157.

3
Organizing

Organizations are dynamic, organization charts are static. Organizations are groups of people, assembled for common purpose. The
chart of their organization is an attempt to delineate the responsibilities and authorities they have or those assigned to the office they
hold. Because of the dynamism of an organization it is impossible to
portray accurately the actual responsibilities assigned or assumed by
individuals and the delegated or actual power they command. If this
is true for the formal organization, think how difficult the portrayal
of the informal organization which exists in every firm.

From a project perspective, organizing means assembling the required talent and expertise and unleashing it to produce the desired
result through judicious allocation of responsibility and authority.
In the process one must recognize the personalities of the individuals
and the form of the parent organization within which the project will
be conducted. A knowledge of how groups function and how individuals react within the framework of group activities is an important
tool of the project manager's trade. Firms in which projects are incidental to the main course of business require a different approach
than firms that are project oriented. The incidental project, being
a temporary phenomenon, can put individuals' temporary and long-
term goals on divergent paths. The most usual form of organization
for such project, the matrix, adds dual loyalties to the equation.

Like the chemist who must combine the right chemicals in the
proper proportions with the necessary heat or pressure, the project
manager must assemble the necessary talent and blend it into a cohesive group to achieve an objective. Doing this requires a knowledge

of group dynamics, the workings of formal and informal organizations and the use and misuse of the formal and informal communications networks.

3.0   GROUP DYNAMICS

What does one organize? People. Why does one organize? To focus
the efforts of people. In the organizing equation the known is the
objective, the unknowns are people. An attempt to understand the
individual and groups in the working environment has progressed
from the scientific management concepts of Frederick W. Taylor,
through the expectancy theory of Victor Vroom and its numerous
variations. Group processes require an understanding of individuals
and their interaction. In addition to their intelligence, individuals
have fundamental needs — for power, affiliation, and achievement.
These needs manifest themselves in personality and action. Various
methods have been developed to determine the strengths of these
needs. In the formation of groups, much attention is paid to the
technical, administrative, or intellectual skills of the members. Little
attention is given to their personality and potential actions based on
individual needs. It is no wonder, then, that many groups fail to
function effectively — not because they lack the skills required by
the task, but because they are unable to interact to put those skills
to effective use.

Every company should, as part of its employee socialization
process, include a program directed at individual assessment and
group interaction. The benefits to the individual and the organization cannot be overstated. The stress, however, has been on the
development of leaders. Leaders are necessary to direct and focus
the effort toward task achievement, but every leader must have followers. There are many more followers than leaders, yet their importance has been unrecognized and their development ignored. A
program to aid individuals to better understand themselves and how
they interact in groups would provide an excellent means to correct
this oversight.*

Expectancy theory attempts to explain individual behavior through
a model linking motivation, effort, and performance to expected outcomes. Tempering performance is the inherent ability of the

---

*One source of assessment of individual values and their organiza-
tional implications can be found in *Career Anchor-Discovering Your
Real Values*, by Edgar H. Schein, published by University Associates, Inc., San Diego, CA 92121.

individual. Motivation is provided by the anticipation that the effort expended will result in a performance leading to anticipated extrinsic and intrinsic rewards. Motivation is self-generated and therefore strongly linked to the values placed on the outcomes and the probability of achieving them. It is at this point that the individual becomes inextricably tied to the environment because it determines the outcome or provides the catalyst leading to rewards.

Performance-based reward systems are the most effective mechanisms to motivate people. Unfortunately, many are poorly structured and inequitably administered. To achieve its desired outcome, an organization must also perform adequately and follow the model by being motivated to expend the effort necessary. This means establishing, with the individual and his supervisor, a mutual understanding of the expected performance requirements and what are the reasonable expectations. This includes the positive as well as the negative. It entails a mutual investment toward increasing the ability to take on different or more difficult tasks and maintaining skills subject to continuous change. Management by objectives (MBO) systems are very effective in implementing this approach if properly administered.

The project group is like an individual in the context of motivation and behavior. It has a behavior pattern and collectively expends an effort in anticipation of reward. The objective is that the individuals on the project team become synergetic and focus their efforts for maximum performance. This entails matching the abilities of the members, their personalities and temperaments, establishing the performance requirements, and setting the standards for rewards. Organization design is the map that should be the result of the process of defining the task, evaluating the human resources, and determining how they can interact to achieve the objective.

3.1   WHY ORGANIZE?

An organization is more than a chart, with little boxes attempting to identify functions, responsiblilities, and the people who fill them. It is a dynamic model of action and interaction. Attempts to depict this can only be a snapshot in time, obsolete the moment taken. Organization is the means by which goals and objectives are allocated to those responsible for achieving them. The formal organization depicts the structure developed and outlines the group responsiblities. Its objective is to divide the effort required in a rational way, avoiding both omission and duplication.

Tushman and Nadler have developed a "Congruence Model" of organizational behavior which has been useful in understanding and diagnosing organizational dynamics [1]. Their model is shown in Figure 3.1. Strategies are developed under the influence of

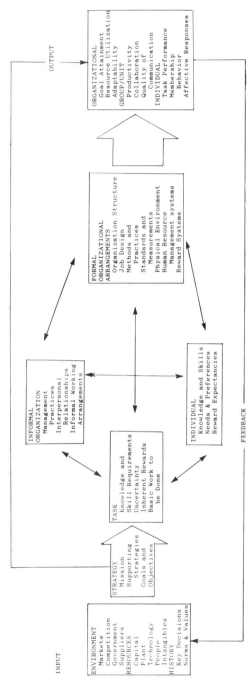

Figure 3.1 A congruence model of organization behavior. (Copyright 1986 by the Regents of the University of California. Reprinted from *California Management Review*, Vol. 28, No. 3. By permission of the Regents.)

environmental factors, available resources, and the traditions and history of the organization. Tasks are outlined that will implement the strategy within a framework of formal and informal structures, keeping in mind individual capability and desires. Organizing for a project is maximizing the congruence between inputs, strategy, elements in the transformation process and, the desired outputs. The following sections are intended to demonstrate how this transformation process takes place in a project environment and how congruence can be achieved.

Organizational requirements are simplified because a project presents a well-defined task on which to focus effort. How this effort is expended will depend on overall corporate strategy. That strategy is open to innovation and imagination, but constrained by the available resources of capital, technology, and people. The project organization must function within the environment created for it. All of the participants on the project should be aware of the goals and objectives of the immediate task and how it contributes to the overall corporate goals.

Functioning effectively as a team requires that all of the members have at least a moderate need for affiliation. Projects demand a singleness of purpose, strong and effective internal communication, and rapid detection and response to problems and opportunities. Independence and self-confidence are important characteristics and ensure the probability that a team member will be a self-starter and require only minimal supervision. Too much independence, however, can manifest itself in poor communication and lack of cooperation. How one fits into the project team will provide a good indication of the fit within the corporate culture.

The project organization is a microcosm. It exists within the larger universe of the corporate organization. To function effectively it must be empathetic. Formalizing the project structure requires care not to depart radically from the framework of the overall organization. If the true operational characteristics of the project organization are to reflect a functional form of organization, then the project manager becomes a facilitator or coordinator. This is an entirely different role than that of one who has direct line authority over the team members and is accountable for project performance.

Rare is the project where a perfect match can be achieved between the individual and the task. The individual may be overqualified, find little challenge, and ultimately underperform. More often he will be underqualified and without help, become frustrated, and also underperform. The objective becomes one of matching to achieve desired performance. This requires artificial initiation of what one hopes will evolve into the informal organization. This will only result from an intimate knowledge of the individuals. That is why it is so important to develop the personal inventory. The good manager is cognizant of this need and significant time is spent getting to know the individual team members and how they interact. It is an active, not a passive, effort. It involves thrusting individuals together to complete a task and obtaining a whole that is greater than the sum of the individual parts.

Project teams have a tendency to perpetuate themselves, particularly if they are successful. The danger is sterility and ultimately failure. The sports world provides evidence of professional teams who were singularly successful, some for many years, but who failed because they were not regularly rejuvenated. In organizations where projects are a way of life, management must be constantly aware of the need to infuse new ideas through new people. It must resist the temptation to try and perpetuate the successful team.

Mixing and matching skills and personalities is a question of function. Communicating it internally and externally is a question of form. The form that the group interrelationships take is organization.

3.2   THE PROJECT ORGANIZATION

The underlying principle of modern organization theory is that different forms of organization are required for different organizational situations....This is the so-called contingency theory of organization, which claims (1) that there is no single, right structure for an organization; (2) that organizational change is often inevitable, necessary and useful; and (3) that subsystems interdependency is especially important to organizational effectiveness. [2, pp. 160–161]

Albert J. Kelley

3.2.1   The Project Organization

Project organizations are, by the nature of their task, functionally oriented. A single function may predominate, or the project may require a composite of skills in order to achieve the objective. The scope of the project will determine the breadth of its organizational structure. Corporate culture will influence the authorities and responsibilities assigned to it. Given these parameters, the project manager can then form his team using the following steps.

1.   Divide the objective into individual tasks. If the project is large, it may be necessary to establish subobjectives to permit greater divisibility. This is commonly called the work breakdown structure. It is important to differentiate between the work breakdown and the structure. Tasks fall into two different categories: those dependent on each

other in a sequential fashion and those which have com-
monality of skills required to perform them.

2. Determine those tasks which have common skill factors, complexity, priority, and performance durations. The result of
this exercise is determination of the number and abilities of
the individuals needed to perform the tasks.

3. Determine the qualifications of the personnel required to perform each of the tasks and note any special skills needed.

4. Examine the whole and make adjustments as required to match
the results of step 3. This may require moving people from
one group to another, combining or breaking down tasks,
and adding or subtracting individuals.

5. Group the individuals in such a way as to reduce the initial
project to one of several projects that minimize dependency
on one another. The maximum of dependency between individuals should occur within their own work group.

6. Supervision is chosen to link the groups through a system
of interdependence which considers skill requirements and
time factors. As with the main project, the groups should
perform their respective tasks, then go out of business.
Continuity can be retained by shifting individuals whose
skills will be required in later groups.

The organization of a project takes place within and is subsid-
iary to the overall organization form. This is true even in organizations that are project driven. Like its parent, the project organization is centered around its objective. The objective is well defined
and restrained in time and budget. Where it fits in the overall organization and the type of organization in which the project is conducted have a significant impact on how the project will be executed
and the style and form necessary to make it succeed.

3.2.2   The Project Within the Organization

The guidelines for organizing a company are beyond the scope of
this text. The number of variables depends on factors such as the
type of business, its size, and its competition. Conversely, a project with a fixed objective, time frame, and budget has sufficient
similarities to permit a nearly universal approach. Projects, however,
must be executed as a part of an existing organization. It is important to know what organizational forms there are and how they may
influence the conduct of the project.

Henry Mintzberg, in *The Structure of Organizations*, has drawn
on the work of many theorists and classifies organizations as falling
into five categories [3]. These are: *Simple Structure*-A centralized organic form most usually associated with owner run and

operated small businesses. Small to medium construction and engineering firms fall into this category. The owner or a few key individuals retain most of the power and authority.

A project will either fall directly under the control of the firms
owner or one of his immediate subordinates. This form of organization is characterized by tight centralized control and management.
It is therefore inconceivable that a project of any consequence would
stray far from the strategic apex of such an organization. There
are fewer formalized processes in such an organization and the project's manager can expect a high degree of autonomy in the routine
conduct of the work, but an equal degree of visibility and perhaps
supervision. It is incorrectly assumed there is more freedom in a
small organization. The converse is more often the case. Smaller
businesses are individually owned and operated. The owners, perhaps rightly, expect to be involved in every facet of it.

Professional Bureaucracy — A conglomeration where the power
and authority are based on highly trained and skilled professionals.
The work is complex, but stable. Hospitals and universities are
good examples of this form. It is decentralized both vertically and
horizontally.

Engineering consultancies may fit this form if its professionals
form only a loose confederation for administrative convenience but
normally deal with their own constituencies. Its primary characteristic is the federation of specific disciplines that require a high level
of training and experience. A greater degree of autonomy is prevalent in the individual discipline groups of such organizations. They
are more decentralized from the standpoint of control. Projects may
be internal to these types of organizations, but they are more often
conducted by the professionals for their clients. Internal or external projects may be undertaken by a single discipline or across discipline lines. If they fall within a single discipline they will be similar to those undertaken in the simple form. In this instance, the
discipline manager substitutes for the owner. In addition, the project will have to conform with the system prescribed by the organization and thus will be more formalized as regards process and
procedures.

There may or not be a project discipline in such an organization.
If one exists, its resources must be drawn from the other disciplines
in order to function. The universal form for the project organization
under these circumstances is the matrix, which will be covered in
greater detail in a subsequent section. The project manager may
have more or less responsibility and authority, depending on whether
the organization is project driven or discipline focused. In either
case, the major mechanism for accomplishing his task is coordination
and liaison with the independent disciplines.

Machine Bureaucracy — Highly centralized technostructure characterized by formalized procedures and processes. Mass production

and process firms fit this category. Some of the very largest construction and engineering firms are in danger of moving in this
direction.

The single most important characteristic of this form of organization is standardization of output. It describes nearly all government organizations and manufacturing enterprises. Ensuring this
standardization are rigidly applied rules and regulations covering
process and behavior. Projects within this framework can be accomplished internally or externally, or by a combination of the two, and
for internal or external consumption. As an example, Boeing utilizes
its own resources and those of subcontractors in developing products
for sale to others. A bank may pull together some of its own resources and hire a consultant to undertake a computerization project. In either case, the project organization should be examined
from the viewpoint of its initiator.

Even in an organization where projects can take on great significance, as with those of Boeing for example, the temporary nature of
a project cycle prescribes a temporary organization. Some form of
the matrix has become almost universal as a means to organize for
these temporary activities. Recognition of the status of the project
within the corporate realm and its dynamic nature determines the reporting level of the project manager and the degree of freedom allowed from normal administrative practices. A project constitutes
change and is therefore dynamic. In order to be successful it cannot be constrained by the same rules that govern and order the
stable environment of normal business. The independence necessary
for adequate execution therefore demands a higher level of responsibility and commensurate reporting for proper administration and control. An unfortunate tendency is to downgrade the status of the
project, particularly at the middle levels of management. This forces
it to conform to administrative and operating procedures established
for normal activities.

Adhocracy — A collective of highly trained and skilled professionals operating in a complex and dynamic environment, usually in a
matrix form. NASA is a classic example of this form. Some consulting and engineering firms operate in this mode. Specialized construction companies may also fall into this category.

The primary difference between this form and the professional
bureaucracy is in the degree and level of coordination between the
professionals necessary to accomplish objectives. No individual discipline is usually sufficient to perform the whole task but each essentially controls its own part. The efforts are usually characterized
by the fact they are of one-of-a-kind, unlikely to be repeated.

This type of organization is primarily project driven. As such,
the project achieves its commensurate status. The pioneering nature
of the effort and the unlikelihood of repetition permit it to escape
the burden of formal discipline. These factors make it easy to

attract the best talent and provide a high level of motivation. The project manager must therefore have broad experience and a high degree of self-discipline to avoid pitfalls that exist due to the absence of rigid control.

NASA's space shuttle program and the Challenger disaster are excellent examples of the benefits of adhocratic project management and its consequences when self-discipline is not present.

Divisionalized Form — A market or geographic decentralization where individual division autonomy functions within prescribed output requirements established by a central hierarchy. The individual divisions may take any of the previous forms, and as a consequence, projects will have the same characteristics.

The reason for considering these as a separate form is that in divisionalization, what may appear as a greater degree of decentralization is, in fact, greater centralization. Although each division now has within it all of the functions to be autonomous, central authority now prescribes the measure of its performance. It determines which divisions will be allocated capital and which will survive or be cast off.

The search for a perfect project organization is a never ending but pointless one. As in all other aspects of project life, perfection is not a valid goal — accomplishment is. [4]

Robert D. Gilbreath

3.2.3   The Matrix Organization

Discussion of project organizations to this point has made little mention of responsibility, reporting, and authority relationships. This has been intentional. Up to now the discussion has centered on structural form. The matrix organization is neither an organization nor a form, but a relationship. An explanation is in order.

A project is an effort that has a clearly defined objective, a time for execution, and a finite budget. The objective can be divided into more or fewer tasks and these assigned to individuals. These individuals may be from a temporarily constituted project organization, or team, be seconded from other organizations within the company, or be from an outside source. Any combination of these is possible. Each individual is responsible for the performance of his assigned tasks and the project manager for the collective tasks and the objective. In the so-called matrix organization, all of the functional personnel, in addition to being responsible to the project manager for the performance of their tasks, are also responsible to the functional organization. This is the condition where an individual wears two hats or serves two masters. These relationships are shown in Figure 3.2. Project participants take functional direction

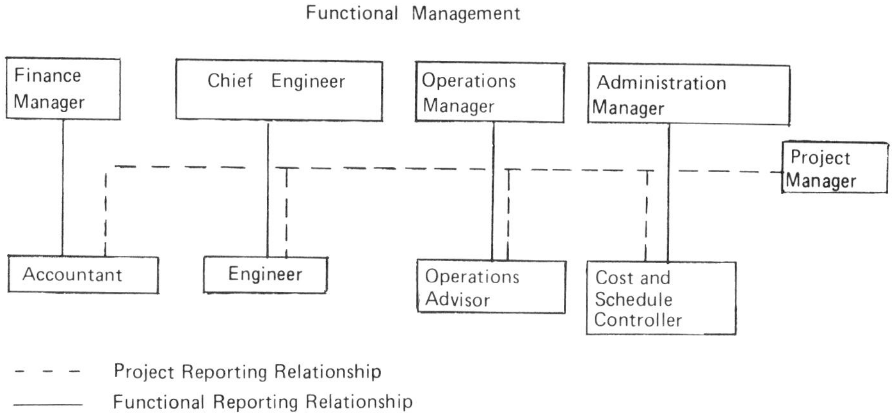

Functional Management

- - -   Project Reporting Relationship
―――――   Functional Reporting Relationship

Figure 3.2   The matrix organization.

from their respective function management and at the same time are responsible to the project manager for their contribution to the project.

The matrix organization is not new. It has existed, without the name, for many years in the construction and engineering industry. Companies in this industry are basically project driven and firms are organized in such a way as to routinely assemble project teams from their ranks. Many of the personnel in the companies that constitute these industries are involved only in projects. They move from one project to another as each project is completed.

It might be said that defense industries, particularly those in aerospace, were responsible for adaptation of the project form which became the matrix. The task of putting together and executing larger and fewer complex projects required a new approach to provide greater efficiency. The response was to second personnel from their normal assignments to a task group with the stipulation they would ultimately return to their regular function. The concept works in some organizations but not so well in others. The reasons for these mixed results are as complex as the relationships inherent in the form.

There are those who consider the matrix a fad whose time has come and passed [5]. It may be helpful to explain why it works in some cases and not in others. These ideas have come from experience in organizations where it has and hasn't worked, but do not necessarily have rigorous research support. As suggested earlier, some organizations are project driven. That is, their survival or at least a significant portion of their business is dependent on the

execution of projects. This fact makes it imperative that the project organization be efficient, responsive, and motivated. To accomplish this, objectives are clearly defined, responsibility assigned,
and commensurate authority delegated. Formalized procedures for
administration have been developed peculiar to projects as opposed
to being imposed from other applications. Where personnel are seconded from functional organizations, the functional managers are
fully supportive of the project concept and the authority of the project manager. Where there is conflict, resolution occurs at most one
level above the project manager.

   The system is founded on trust —trust that the project manager
is intelligent, experienced, and has enough good sense to seek out
expert help in areas where his own expertise is limited, trust that
he will properly exercise the authority necessary to execute the
project in an efficient manner and in conformity with corporate policy. When this trust is absent, controls are imposed which effectively castrate the project manager. The result is to inhibit progress,
increase the generation of paper, and give vent to disruptive im-
pulses of functional subordinates. Worse, controls are minimal, authority diluted, and the project manager left to flounder or operate
in a high-risk environment. This is a no-win situation for the project manager and a no-lose one for the functional managers.

> The cultural characteristic of the matrix design causes
> two key attitudes to emerge: the manager who realizes that
> authority has its limits and the professional who recognizes
> that authority has its place. [2, p. 172]
>
>   Albert J. Kelley

   The absence of autonomy places a greater emphasis on the interpersonal skills of a project manager. This is not to say that these
skills will not serve any project manager well. There are times,
however, when decisive action call for a more dictatorial style, and
if authority is absent, flexibility to adjust to the demands of the moment is also absent. When functional authority is predominant, the
project manager's status is reduced to that of a beggar.

   There are two areas where efforts can be made that will improve
the functioning of the matrix organization: fostering a higher degree of trust and recognition of the needs of the individual. Many
people unfortunately feel they have a corner on integrity and if they
were not watching, others would give away the store. This is
"Theory X" mentality. It is one of the reasons there are so many
layers in American and European organizations. Close supervision
and narrow spans of control offer some comfort when trust is missing. Middle Eastern organizations in many instances go to the opposite extreme, where the owner trusts no one. This organization

structure is flat, and as it increases in size, it becomes totally un-
manageable under such a philosophy. One of the secrets of the
Japanese success in making the flat organization work is that there
is an understanding of the other fellow's job and his place in the
scheme of things. Each has the same objectives and is sympathetic
rather than suspicious [6].

Employees assigned to a project have a legitimate concern as to
their status and future. If the individual and the company are to
gain the maximum from a project assignment, it must be acknowledged by both that the employee can benefit from the assignment
and that such benefits will depend on performance. Project assignments are an opportunity, and to be selected is recognition of prior
performance. This can only be accomplished with full cooperation of
the department from which the individual is seconded. If a formalized individual development program is in place, this becomes a much
easier task. Corporate management should encourage an environment
where functional managers fully support the matrix concept and
their own futures depend on the performance of their personnel on
project assignments. Performance reviews and proposals for salary
increases or promotion should be a joint effort of the individual's
project supervisor and his functional counterpart. The performance
interview should be jointly conducted with the employee.

Peter Drucker has surmised an attitude that provides one of the
keys to make the matrix succeed. "A man who knows that his promotion depends entirely on the powers that run the accounting department, will emphasize 'professional accounting' rather than contribution to the company. He will have a greater stake in the expansion of the accounting department than in the growth of the company" [7]. Changing this perception is a responsibility of management. If it is not accomplished, a project manager's lot will indeed
be unhappy.

3.3   THE INFORMAL ORGANIZATION

3.3.1   General Description

Have you ever wondered how a task that seemed impossible was accomplished so quickly or how news got out that the board has just
replaced the president? Meet the informal organization. No one
disputes the existence of the informal organization. Call it the
"buddy network," the "old boy club," or what you will. As long as
people are engaged in common effort, the informal organization will
exist. If it doesn't, it will be created. It exists because formal
organizations are imperfect. There are gaps and overlaps which
must be broached or bypassed to overcome inherent deficiencies.
Communications do not always satisfy the desire to know, or the
speed with which knowledge is wanted.

> In the process of developing a formal structure...the
> aim is to produce a system that will function as planned
> and to ensure that any informal organizations that develop
> within it will not function in such a way as to make attain-
> ment of the formal objectives less likely. [8, p. 180]

Ernest Dale

Like the formal organization, the informal organization has its
objectives. These are information and action. They may be, and
usually are, mutually exclusive, in that the information organization
may be entirely separate from the action organization. Speculation
into how these organizations function should provide insights on how
to make them work in a positive way and how to tap them for
results.

As its name implies, the informal organization functions within
the formal organization and in parallel with it. Its objectives are,
for the most part, complementary — at least those which should be
encouraged. The informal organization also has its reward-and-
punishment system. How does one get to be part of this organization, how does it function, and what is to be gained or lost as a
result of being part of it?

Nearly everyone is a part of some informal organization for three
reasons:

1. A need for affiliation.
2. The position one holds enables others to accomplish things
    they would otherwise have difficulty doing or find impossible.
3. Referent and expert power inspire confidence and attract
    others.

The organization places individuals in a group environment not
of their own making. It is unlikely that any group can succeed in
its objectives without the help and assistance of others in the organization. Finding and cultivating these associations results in development of the "action network."

3.3.2   The Action Network

At each successive level in the corporate hierarchy, accomplishment
should increase through the magnifying effect of the number of subordinates available. There is a greater authority to establish priorities and set deadlines. Interdependencies within the organization
require periodic adjustments to these priorities and deadlines. Re-
ciprocal concessions cement the bonds of these interdependencies.
The further up in the hierarchy, the stronger the bond and the
more power invested in the informal organization. The channels to
get things done change, as do the things themselves. We are now

able to do more through the route of formal power. It is not always true that favors are done in the hope they will be reciprocated. There are certain individuals for whom things are done simply because they are admired and respected. These individuals are more visible in the fields of religion and politics, but they certainly exist in the business world as well. They are said to have charisma and can be found at all levels. There are also those who take pride in accomplishment and recognition of effort. Reciprocity need not be expected, thanks constitute a sufficient reward.

The action network is inherently complex, more so than the formal one. The reason is obvious when one recognizes that as people move up or out, the entire structure has to be rearranged. People who move into the position vacated are not automatically part of the old order. At each level, exploration is needed to find and develop the new network. There are also times, perhaps when the well is tapped too often, or a commitment not kept, that the norm of the informal organization is broken. Rejection by the group can be the result.

3.3.3   The Communication Network

The communication network, or grapevine, functions in a different manner and includes different people, although there may be some duplication. It is also selective, in that participants in the network vary depending on the type of communication. The flow within the network moves in both directions, but is primarily from the top down. Communication is almost always oral.

There are several reasons for communicating information that is not widely known. First, to make it known that the communicator is part of the inner circle and privy to information others are not. Second, to pass it along to others without necessarily divulging the source or to keep the channel open. Third, a desire not to put something in writing, but to ensure wide dissemination. Selectivity of the listeners and free communications ensure participation. Like name dropping, passing along information not widely known is a way to boost one's ego. As a consequence, the type of information one has that will achieve this objective is communicated to selected peers and, occasionally, subordinates. It is done to elevate one's status. This type of information may or may not get passed on, since dissemination was not the primary intent.

There is selectivity with information passed on for further dissemination. It must be communicated to someone who will pass it on and make sure that the audience to be reached is the one intended. There is only one way to determine this: by trial and error. It is important to remember that as you believe you have developed a network of communications, you are also part of one. Often, as part of the obligation of participation in the action network, we exchange

information when favors cannot be replaced in any other form.   In
this case the two networks tend to overlap.

3.3.4   Making It Work Positively

Moreover, the informal organization has no stability.
Alliances are formed and broken; new conflicts arise within it; a change in the incumbent of a single executive job
can change its character entirely. No company could hope
to develop and reach clear-cut goals if it built its organization structure on the shifting sands of the informal
organization. [8, p. 287]

Ernest Dale

If the formal structure is not to be adjusted to incorporate the informal, how can the apparent efficiencies of the informal one be gained?
One way is to examine how and why the networks form, who belongs to
them, and then make appropriate adjustments. An actual experience
will give you a good idea as to how this can be accomplished.

Take the case of an actual company, structured along functional
lines.   Projects were carried out by a permanent organization within
the engineering function.   The project managers were responsible
for the total project, from design to startup of the facilities.   Procurement of materials and equipment, although part of the overall
project responsibility, was performed by personnel in a completely
separate function.   This function's primary responsibility was routine procurement of an array of common materials needed to maintain
company operations.   It was never organized to procure massive
amounts of one-of-a-kind, one-time items.   In essence, the project
prepared the necessary paperwork for procurement, turned them
over to the procurement agency, and waited for the items to be de-
livered in conformance with the provided specifications, quantity,
and delivery date.   In addition to standard bread-and-butter items,
there were others of complex highly specialized machinery with detailed technical specifications.

The foregoing represents a classic case of responsibility without
authority.   Nevertheless, this situation existed, was acknowledged
by corporate management, and was retained.   Without any accountability for project purchasing and their normal accountable functions
to perform, project purchasing had the obvious lower priority.   The
expected problems of delay, failure to meet specification, and shortages or overages developed.   The action network came into being,
but because of the major differences in functional responsibility,
there were few individuals who even knew each other or had much to
exchange.   Those few who did have a common link were at the middle-
management level owing to a concurrent expansion of the company.

One major change that improved the situation considerably was physical relocation of the purchasing function representatives to the project offices. This accomplished two objectives: it cut down the channel of communication and, through the now present representative, added his action network to the project.

This example has not been lost on many organizations, particularly those in the engineering and construction industry. Physically locating project teams and their supporting organizations improves communications and fosters development of positive-action networks.

The informal organization exists or it will develop. Its benefit is the ability to smooth out the rough spots in the formal organization. Although the formal organization is intended to reflect things the way they are, it seldom does, but should come close. If it does not, and informal action must span several levels in the organization for it to be effective, it is time to examine the formal structure. Too often structural change is an accommodation as opposed to the result of careful analysis of need. The action network adjusts, but the out-of-tune formal structure creates roadblocks.

Building action networks should be part of the formal training process. Not that it should become formalized in its own right, but the organization must acknowledge that one of the benefits of rotational assignments within it results in the building of friendships and alliances which will become valuable. Such assignments must therefore be of sufficient duration. Three to six months is not enough, and the time should be as much as is necessary to get to know the job and to make some contribution. It is difficult to build alliances if the assignment is all taking and no giving. The broader and more varied these development assignments, the greater the benefit to the individual and to the company.

Unless a company is prepared to install a public address system and broadcast its information, there is no more efficient communication network than the grapevine. It can, however, distort true information and create information that is untrue. Although some of the negative aspects of the grapevine can never be completely obliterated, two basic rules will minimize them. First, distribute information frequently, widely, and promptly. Second, don't try to keep secrets. Easier said than done. An open communication system, which provides more than just company propaganda, is essential to a high level of morale. It will reflect itself in performance.

The communication network should be a two-way system. The prudent manager knows how to tap the network to get the pulse and substance of what is actually going on. It can provide early warning signals of situations or conditions that require attention. If results are indeed forthcoming, one will find that he is permanently plugged into the network.

No formal information system, including all of the most elaborate management information systems (MIS), will ever substitute for the

communications network. The reason for this is that the most valuable, or "soft," information never gets communicated in a formal way.
An executive, manager, or supervisor will soon be out of touch with
his organization if he cannot tap the information network. He must
also make direct contact with his subordinates and associates and
listen to what is happening.

3.4   AUTHORITY, RESPONSIBILITY, ACCOUNTABLILITY, AND DELEGATION

An activity is not a project until top management
thinks it is. Only when senior managers controlling the
resources necessary to accomplish the assignment are
willing to consider it a project and make the necessary
delegations of authority is there likely to be a useful
application of project management. [9, p. 2]

Charles Martin

3.4.1   Definition

There are many definitions of the terms authority, responsibility,
accountability, and delegation. Taken together, in the context in
which they will be treated, they are; the right to act, the obligation to act, the liability to act, and the duty to get others to act.

3.4.2   Authority

The right to act is explicit in the position one holds. This is usually codified in documents, such as general instructions, powers of
attorney, and the like. This right emanates from the articles of incorporation and the bylaws that establish the business. They permit
its officers to act in pursuit of its objectives. The board of directors is responsible to the stockholders; the officers of the corporation, to the board; and employees delegated authority, to the officers. With these rights come obligations. Members of the board of
directors have fiduciary obligations to the shareholders. The members act on behalf of the stockholders and there is a recognition in
law of the trust implied in this relationship. If that trust is betray-
ed, the board members are liable, collectively or individually, to
legal action. A similar relationship of trust exists between the board
and the company officers and between the officers and their subordinates. Although legal action or removal by failure to reelect is
not the usual mechanism for the failure of corporate officers or their
subordinates, termination certainly is.

How the various members of the corporate hierarchy exercise
their authority and discharge their responsibilities provides an

answer to why one never has the level of authority to go with responsibility. The corporate board has the responsibility for all corporate
action, including capital expenditures to cover projects. Some of this
responsibility is delegated to the chief executive officer (CEO) in the
form of broad exercise powers within approved limits. The CEO further delegates parts of his responsibilities to elected corporate officers
for actions within their respective areas. This could include a divisional vice president who has authority over budgets and capital expenditures, within his approved limits. This authority may be further
delegated in regard to capital project expenditures for facilities to the
manager of the engineering department. Through him, the project
manager is responsible for his assigned project. In order to maintain
control and as a reflection of the degree of responsibility assigned each
level, the authority to commit the organization in the form of monetary
limits is imposed.

Another aspect of delegation is the relationship with one's supervisor. It is usually a relationship developed over time and is not anywhere near the formal nature of the official delegation. Figure 3.3
illustrates how this delegation functions.

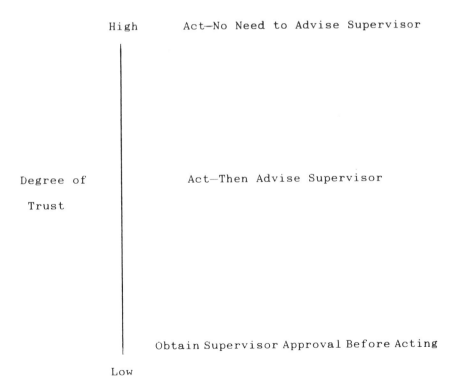

Figure 3.3   Superior/subordinate delegation.

As confidence in a subordinate builds, prior approval will diminish and the next levels increase. This relationship covers not only the official authority of the subordinate, but that of the supervisor. Where there is a high degree of trust and reliability, the subordinate may have nearly the same authority as the supervisor. Where this is absent, the subordinate may need to seek approval of the supervisor for nearly every decision he makes, including those which have been officially delegated. Fostering this relationship toward the highest level of trust and reliability should be the objective of both the supervisor and subordinate. It is of mutual benefit to both and an enhancement to their success.

3.4.3 Responsibility

As a project manager, you will be responsible for the execution of your project. Your supervisor, who may be a corporate officer, is also responsible for the execution of that project. The board of directors is responsible for all corporate activities to the stockholders, and that includes your project. Each level in the corporation therefore retains responsibility for all of the actions it has delegated below it. A better way of visualizing this responsibility is to divide the corporate pyramid vertically, as opposed to horizontally. This is illustrated in Figure 3.4.

Many of the complaints of managers about authority commensurate with responsibility are well founded. This is despite an understanding of the principles of ultimate responsibility. These complaints originate not from lack of authority per se, but what is delegated and how.

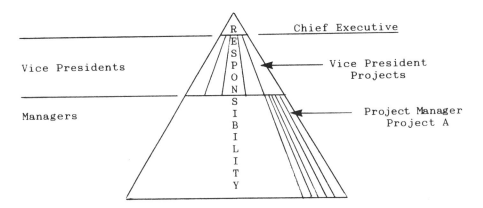

Figure 3.4   The responsibility pyramid.

3.4.4   Accountability

Accountability follows naturally from authority and responsibility. It is the concept of being answerable for management's trust in the individual. The actions of each member of the project team stem from specific and implied authority and their perception of responsibility. Confirmation and reinforcement of trust come from the proper exercise of authority and execution of responsibility. Accountability is also dependent on authority and responsibility. If trust is absent, authority and responsibility will be reduced and their exercise and execution will produce a lesser result. If these factors are not commensurate, the expectations of superior and subordinate will diverge, and this may have serious consequences.

3.4.5   Delegation

Three cases of delegation, which encompass nearly all conditions, will be examined. An assumption will be made that all decisions made on the project are those which would have been made by the project manager. If made by others, they are those which the project manager recommended or with which he concurs. The validity of this assumption will be discussed later, so as not to distort the presentation at this point. If the project manager was delegated full authority to execute a project, consultation with others would be voluntary, not mandatory. The project could proceed at whatever pace, with whatever action, he deemed appropriate to the budget and time schedule. As requirements are introduced to review and approve actions prior to their being taken, the overall project schedule is affected. This is particularly true for activities that are critical to the schedule. This is illustrated in the top three sections of Figure 3.5. Responsibility for the actions of subordinates, proximity, and familiarity assist in minimizing the time required by supervisory reviews and approvals. Such is not the case when actions require the review and approval of other parties or agencies. Lack of responsibility and familiarity plus distance lower the priority of action by others. Coupled with supervisory reviews and approvals, these outside needs maximize the time that must be allocated for these processes. This is not to suggest that critical actions should not come under the scrutiny of immediate supervision and outside agencies and parties impacted by the decision process.

Procedures should be established to integrate this scrutiny within the process and minimize its serial impact on the schedule. This is illustrated in Figure 3.5, where supervisory and outside reviews are on a level with project action. Two criteria must be followed in designing such a process.

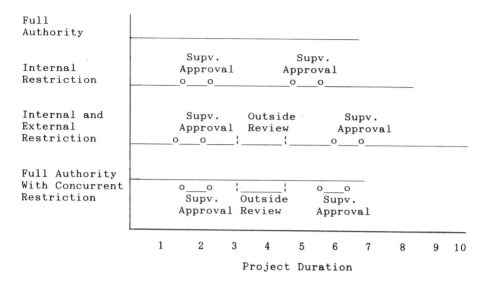

Figure 3.5   Project duration under various forms of delegation.

1. To the extent possible, actions requiring such treatment should be minimized and identified in advance.
2. Actions taken during the review and approval process should not be irreversible.

The objective is to give those charged with review and approval of action an opportunity to input the decision process, while at the same time not making the actions of the project manager a fait accompli. Knowledge that action may be taken during the process will impose a priority on external agencies and tend to reduce the time taken for review and approval. The requirement that actions taken should not be irreversible will encourage the project manager to communicate more effectively with supervision and external parties in order to propose actions that have been developed in cooperation with the respective parties.

3.4.6   Management Overview (Controlling Delegation)

Before discussing techniques to minimize the negative aspects of management overview of project activities, there is a need to clear up any concern about the assumption made earlier. That assumption concerned decisions by the project manager. The majority of all decisions concerning a project are made by the project manager or one of his subordinates. Even in cases where decisions are reserved to

higher levels of management, the decisions will usually come from
alternatives and recommendations made by the project's management.
If the project manager has done his job thoroughly and presented
his arguments in a logical and straightforward manner, his recommendations will usually be accepted. Decisions the project manager
cannot support are so infrequent that the original assumption can be
safely made.

Recognizing that management has a responsibility for project
overview, what steps can be taken to minimize the cost and time impact of the required inputs? There are several:

1. Timely and thorough planning
2. Management by exception
3. Early and parallel development for decision points
4. Elimination of nonresponsible decision points

There is really no substitute for thorough planning. Decision
points can be identified and alternatives evaluated. If management
must make decisions, they can be made during the planning cycle as
opposed to the implementation cycle. This plan might be called a
project execution plan and can be formalized and approved in advance by the appropriate parties.

An effective time saver for decision makers is the flexibility of a
management-by-exception system. In such a system the decision is
made at a lower level, communicated upward, and, after a specified
time, implemented unless exception is taken. As a supplement to
this system, control is established within a specified range of the
desired results. So long as the results achieved are within the acceptable range, consultation or review is not mandatory.

If there is no substitute for higher management review and decision making, preparations and the actual decision-making process
should start early. It should be in parallel with normal project execution, as illustrated in Figure 3.5. The additional cost associated
with the preparation and presentation of information for management
decisions has not been eliminated, but the time impact has.

It is a project manager's unhappy lot if he must obtain decisions
regarding his project from persons who are neither responsible nor
held accountable for them. Elimination of this condition is the only
effective solution. Experience has shown that where this exists, the
corporate culture has condoned it. Having a go at change may still
bring some positive result, and if that fails, one can proceed with
an early and parallel effort, as previously described. Frequent ex-
pediting of the decision makers is strongly recommended.

Project management is a unique endeavor in that its normal scope
encompasses a broad spectrum of functional activities. These may be
research, purchasing, personnel, engineering, accounting, operations, manufacturing, and the like. In order to function effectively
across such a broad spectrum, a project manager must be experienced

and have broad authority. It is not unusual for an individual with this breadth and depth of scope and authority to report at a high level in the organization. When this authority is not present and the reporting level is not commensurate with the responsibility apparent in the task, problems will naturally develop.

> Project Management is a weapon that can be brought out of the arsenal, fired with great success and then put away until needed again. It is a weapon worthy of consideration in all urgent situations. [9, p. 187]
>
> Charles Martin

Many organizations, particularly those which are not project driven, fail to recognize that the benefits of project management do not come without a price. The price of the inherent efficiency and the single responsibility is delegation and control (see Fig. 3.6).

In Case A, the project manager reports directly to the president. Although he may have to obtain his resources from the functions under control of the senior vice presidents, he has equivalent status and authority. Resolution of differences is only one level away. This situation recognizes that the project manager's responsibility encompasses some or all of the functions under the control of the individual senior vice presidents. Authority is necessary within the limits of the project's scope and corporate policy with regard to

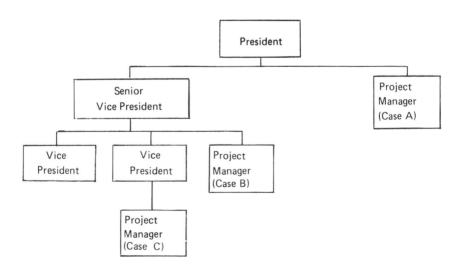

Figure 3.6   Various levels at which project managers report.

those functions. This may appear to be a rather high-level position
for a project manager, but if one considers the scope, size, and importance of some projects, it is not unusual.

In Case B, the project manager is now reporting to one of the
senior vice presidents. This is appropriate if the project does not
involve functions under the other senior vice president. When this
is the case, the resultant relationship violates one of our primary
rules: resolution of conflict should be at most one level higher. In
this instance, if the project manager has a conflict with his counterpart in another function, he must have his senior vice president in-
tercede and, if that is unsuccessful, must get the president to resolve it.

Case C is but a repetition of the second with similar problems
occurring within the project manager's own function. This reporting
relationship can be taken further down into the organization with the
problems becoming more complex. This situation is not uncommon.
There are cases in which project managers are reporting through as
many as four levels of functional management. Correction of such a
situation is obviously beyond the scope and capacity of an individual
project manager. It is the consequence of an attempt to obtain the
benefits of project management while attempting to impose controls
whose negative impact was the basis for implementing it in the first
place.

A manager must obtain results through others. If they are his
subordinates, he has the formal power to get things accomplished.
He will be more successful if they have afforded him referent power
as well. If they are his peers, he can tap the formal power of their
common superior. If not, he must rely on his interpersonal skills,
or perhaps his action network. Having to accomplish a task when
none of these factors are present places an almost impossible burden
on a project manager. It would demand an exceptional level of interpersonal skill very few possess.

3.5   SUMMARY

The organizer must keep in mind that what he is organizing is
people — people who have their own personalities and motivations and,
when assembled in groups not of their own choosing, function in
many different and sometimes unpredictable ways. Organization is
the matching of individual tasks to the individuals in such a way that
they are complementary. When taken together, they form a cohesive
whole. The resultant grouping can be assembled in a functional
format where the tasks have a common skill basis, such as engineering, finance, marketing, and so on. They may also be assembled by
output, or product, such that all the skills necessary to produce the
result are included. Modifications of these are a form of staff

organization along functional lines and an operational organization along product lines.

The project organization exists within the other forms and includes all of the functions necessary to execute the project. The matrix organization is not an organization but a relationship. Personnel from various functions are seconded to a task force or project and have a dual reporting relationship. They are responsible to their functional management and to the task force or project management. The informal organization consists of the action network, which represents the way things really get done, and the communications network, commonly called the "grapevine." Authority is the right to act, responsibility the obligation to act, and delegation the duty to get others to act. No one ever has the full authority to act since those who are responsible for those actions reserve authority over certain actions for control and the exercise of their own responsibilities. The linkage of authority, responsibility, and delegation can be assembled in such a way as to be both effective and efficient. The successful organization reflects the need to develop a harmony between the corporate environment, its strategy, and the transformation process necessary to produce results. The transformation process is that of combining the task, the individual, and the formal and informal organization to produce synergy.

REFERENCES

1. Michael Tushman and David Nadler, Organizing for innovation, *California Management Rev.* 27(3), 74 – 92 (Spring 1986).
2. Albert J. Kelley, *New Dimensions in Project Management*, D. C. Heath & Co., Lexington, MA, pp. 160 – 161.
3. Henry Mintzberg, *The Structure of Organizations*, Prentice-Hall, Englewood Cliffs, NJ, 1979, 512 pp.
4. Robert D. Gilbreath, *Winning at Project Management*, John Wiley & Sons, New York, 1986, p. 68.
5. Robert Youker, Is the matrix fad a fast fading flower? Proceeding of the Project Management Institute Seminar/Symposium, Philadelphia, PA, Oct. 8 – 10, 1984.
6. Trust: The new ingredient in management, *Business Week*, pp. 62 – 63 (July 6, 1981).
7. Peter F. Drucker, *The Practice of Management*, Harper & Row, New York, 1954, p. 223.
8. Earnest Dale, *Management Theory and Practice*, 2nd ed., McGraw-Hill, New York, 1969.
9. Charles Martin, *Project Management*, AMACOM, New York, 1976.

4

Staffing

Organizations are more than blocks on a chart. They are people, deliberately chosen, fitted into a designed interrelationship, and nurtured to produce an established goal. This deliberation requires a knowledge of the task, the skills of the individuals, and the consequences of interrelating a variety of personalities.

The short duration of a project places unusual demands on the selection of staff. People must rapidly climb the learning curve, adjust quickly to each other, and develop synergy. Exacerbating the process is the fact that often, the assembled personnel have never worked together before and most are unknown to each other. This places a premium on proper evaluation of interpersonnel skills of the team members, particularly those of the key leaders of the group. The importance of this process is confirmed by the actions of project-driven organizations in trying to maintain cohesive and effective project teams.

The careful design of interrelationships and selection of individuals to form them can all be for naught if the team members cannot see beyond the completion of the project. Selection to work on a project must be an interim step in the career of each individual and the individual made to feel a part of the development process.

Staffing for projects often involves engagement of outside firms and individuals to supplement the efforts of in-house resources. An equal degree of care must be placed in their selection and the allocation of tasks and responsibilities to avoid duplication or omission. Picking a contractor or consultant may be even more important, as the consequences of a poor choice can be more severe and less easily corrected. How the choice is made may preclude very desirable alternatives and options.

4.0 SELECTING THE PROJECT MANAGER

> Good people alone cannot guarantee project success; a
> project that is poorly conceived, badly planned or pro-
> vided with inadequate resources has little hope for suc-
> cess. But even if there is only marginal hope, an ex-
> cellent team will be far more likely to bring the project
> off than a mediocre one. On the other hand, if the team
> is not up to the job, no amount of brilliant planning and
> lavish use of resources will save the situation; elaborate
> control systems will only highlight the inevitable and
> accelerating approach of doom. [1]
>
> Charles Martin

4.0.1 Picking the Man

A project may consist of many smaller projects, each executed by a different company. It is not uncommon to have several project managers on a single project. Each company, which has a part of the overall, considers its part a project, and therefore, its leader will be called that company's project manager. The concepts and principles that follow apply to any of these project managers.

Job descriptions of project managers no doubt exist in all companies that execute projects. There is no one person to whom they fit. Preparers of these job descriptions believe that managing a project requires someone with the technical expertise of an Albert Einstein, the financial acumen of a Bernard Baruch, and the leadership of an Alexander the Great. Knowing dozens of successful project managers, I can attest this is certainly not the case. What causes this apparent contradiction?

A project, however small, is a highly visible effort. There is a strong desire for it to succeed. Writing a job description that all but those capable of walking on water would find it difficult to fill is one way of trying to ensure success. It also provides an escape hatch for some, who could point to the difficult in finding someone qualified, if the project turns out badly. The tough-to-fill job description does satisfy two basic needs. The requirements almost assure that when the job evaluators get through evaluating the position, the salary range level of the job will likely attract well-qualified applicants. If anyone comes even close to meeting the requirements, he should be adequate to the task. Setting the requirements and filling the job should not, however, be that difficult.

A project manager is first and foremost a manager. The mistakes made in selecting a person to manage a project are similar to those associated with picking any manager of a functional group, the major one being that of selecting the best technician and finding out that not only have you lost a good technician, but now you have a bad manager. There is nothing wrong with a manager having excellent technical skills. Such ability will be helpful in evaluating the work of his subordinates and in gaining their respect as well as that of his peers. Sooner or later one has to make a choice since upward mobility through either route demands full-time dedication. If the choice is management, he will soon fall behind his subordinates in technical competence.

There are three transitional levels in the corporate pyramid: worker to supervisor, supervisor to manager, and manager to executive. None is more difficult than that of supervisor to manager. It is at that point that the umbilical cord to the specialty is cut. There is little hope of a successful return. It also means that to continue to climb the corporate ladder, the previous job must be left behind and the new one hung onto tenaciously. Retaining the old job as a crutch is one of the major causes of managerial failure. A successful manager is one who has navigated this transition.

If you cannot maintain a competence in your specialty and cannot continue to do your old job, at what do you become competent? One thing obviously has to be delegation. The first job is to get someone to perform the old function. Another is to get others to do the myriad of things which need doing and which are necessary to achieve the objective. In the process, the key elements of each of these areas are learned sufficiently in order to pull them all together and direct them to a common goal. There is no better parallel than that of the orchestra conductor. He may be a competent musician in his own right, but probably the first to admit that he is less an individual artist than the orchestra's best. He has an excellent ear for music and can tell when any of his players is not in harmony. Most of all, he has a grasp for the piece and a sense of the musical message the composer was trying to communicate. His musicians know and respect this and will follow his lead.

The picture of our project manager is now coming into focus. He is a manager. He is competent, but probably no longer expert in a given specialty area. His knowledge now tends to the general, but it is sufficient to the task. He knows how to plan and divide tasks and to delegate them to others and monitor their performance. Others look to him for direction, guidance, and help and will follow his lead. Now that we have determined who we are looking for, where do we find him?

Obviously, if a company has current and continuous project activities, there is a cadre of experienced project managers at hand. Moreover, the active projects provide the testing, proving, and development grounds for new project managers. Most do not have this luxury and are engaging in project management for the first time. They execute projects infrequently and do not have permanent employees assigned to this function. These organizations have particular problems.

For the one-time project operation, the primary problem is getting the most qualified individual to accept the job. The project is usually confined to a narrow area, such as a computer application project, modernizing a production line or moving to a new location. One obvious candidate to manage such a project is a manager from the function most influenced by the project or around whose specialty the project has evolved. The individual is a proven manager and has an interest in the outcome of the project. He can probably count on the support of his peers in other involved functions through prior contact on committees and the like. The project assignment may, however, be seen as a sideways move with little future. Another is that the old job will have to be filled and may not be there when the project is over. These fears are real and must be eliminated. Avoiding them by making a second-best choice is to be resisted. If the project involves a new operation, plant, or facility, this may be a promotional opportunity. The job can be put forward as recognition for a good performance. Opportunities available to the manager in his former function should be maintained and his chances for promotion enhanced as a result of taking on the project. If the only option available is a return to the old assignment, it must be recognized and the experience on the project should count heavily toward future development opportunities.

An alternative choice is to bring in an experienced manager from another part of the company. Some of the same problems mentioned above must be faced. In addition, some resentment may be created by those who believe the opportunity should have gone to someone from their own area. Another disadvantage is the new manager's lack of an informal action network. Such a disadvantage should not be minimized. Project durations are usually short, and by the time the new man develops one, if he can amid the possible hostility, the project may be over. Both these problems can be avoided by making an alternative choice from within the organization. There will be less difficulty with his peers, since they may already be known. A new action network can be built rapidly from segments of the former. Any lack of experience and functional knowledge can be augmented by careful selection of support personnel and closer management overview.

Bringing in a project manager from outside the company should be the last recourse. Resentment from within may never be overcome. Lack of any informal network and unfamiliarity with any of his new peers is a large initial burden. If this is the only alternative, the selection process must stress finding someone with exceptional interpersonal skills and the socialization process must emphasize frequent and visible support from top management.

4.0.2   Matching the Man and the Job

Where to search for the right man to manage a project depends to a great extent on the job itself. The job specifications should determine, or at least influence, the functional expertise of the potential manager. The involvement of company organizations in the execution of the project may determine from which function the manager originates. The extent to which outside consultants or contractors are used will also influence the choice.

The scope of a project will determine the degree of special expertise or experience of its manager. If it is a highly concentrated functional project, for example the upgrading of a computer system for a bank, the project manager should have an intimate knowledge of the bank's accounting systems. Computer expertise may also be necessary, but without knowledge of the application, the project may be lacking adequate direction. On the other hand, a project that involves building a new facility demands a much broader knowledge. The scope of such a project is varied and experience in the objective of the project, such as a chemical process, may be necessary, but experience with engineering design, material procurement, contracting, and construction is also required. If any or all of these are missing, reliance on the experience of other complicates management of the project and places further burdens on the manager.

Who will execute the project also affects the qualifications of its manager. There are three modes of project execution. The project can be executed by in-house resources, by outside resources, or a combination of the two. The latter can range over a wide spectrum. With only company personnel involved in project execution, coordination is easier due primarily to singleness of purpose and a base of operation familiar to all participants.

Having a project executed by an outsider poses unique problems. The first is selecting someone qualified to carry it out. The second is ensuring singleness of objective and purpose. A third is evaluating the performance of the project. The most important is the recognition that someone has been chosen as a proxy, for whatever reason, and that he be allowed to perform as such. A client once told me that my company was engaged to handle his

project because we were the best in the field and his firm had no
expertise in it to do the job themselves. I have never forgotten
his advice to himself and others in his company regarding the project's execution. He said "Don't buy a dog if you intend to bark
yourself." The selection of a surrogate project manager and company will be covered in greater detail later. The point to be made
is that the company representative, or if you will, project manager,
chosen to oversee such an arrangement requires unique skills. He
must be able to evaluate performance and progress from reports
that may be biased and contact with but a few people of the company hired to undertake the work. He can be helped by an effective control system, but he must have excellent analytical skills
and be able to zero in on the real problem.

Many projects require utilization of a combination of company
as well as outside resources. The degree and level at which these
combinations apply will vary widely. The challenge to the project
manager is to effectively coordinate the efforts of both groups and
to integrate personnel into a functioning team. This has to be accomplished with the minimum of conflict and duplication. Accomplishing this requires a manager with sufficient authority and a high
level of interpersonal skills. This type of project puts a premium
on the qualities of a generalist as opposed to a specialist. The
manager should also have a well-developed action network.

Figure 4.1 is a nomograph illustrating the skill and experience
qualifications of the project manager. In the example shown, the

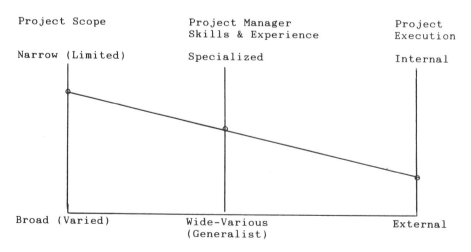

Figure 4.1   Project manager skills and experience selection nomograph.

project scope is relatively narrow, but will be executed using mainly external resources. Connecting these two points crosses the
project manager skill and experience line near the midpoint between
that of a specialist and a generalist. This is reasonable in that
the narrow scope of the project puts a premium on the manager's
technical skills, while execution requires experience in dealing with
outside agencies, primarily engineering and construction contractors. It may also require dealing with the public and the government and interfacing all of these with internal departments. The
experience, qualifications, and skills thus far described as needed
by a project manager will narrow the choice. This in itself is helpful in providing appropriate recognition of the importance of the
job. For the infrequent project operation, this may be the stimulus to those reluctant to become involved. It is a reinforcement of
the importance placed on the project concept for those who regularly apply project management.

4.1   MANPOWER SOURCES

An army consisting only of a general is unlikely to meet with much
success. So too is a project with only project managers. Helpers
are needed, who can range from a few part-time assistants to a
permanent cadre of thousands. Where this help will come from presents both a challenge and an opportunity. It is a challenge in
that use of internal resources may dilute efforts on other objectives
or raise an issue of underutilization or may be an opportunity where
personnel can be developed and given additional experience. It is
a challenge in that the use of outside resources presents a wide
range of unknowns, or an opportunity to the extent the company
can accomplish the project with outside help and still achieve its
other objectives. Next to selecting the project manager, no other
decision will have greater influence on determining the success of
a project.

4.1.1   Internal Sources

First consideration must be given to selecting project personnel
from sources within the company. These personnel bring certain
assets that an outside agency can never begin to match. Important are company and project identification, knowledge of company
systems and procedures, a sense of permanency, and established
action and communication networks. Associated with this selection
are the same problems faced when picking the manager. What incentives are there in opting to take a project assignment and where
does one go when the project is completed? If the project crosses

functional lines, who leads the effort and how do the other functions interrelate? Staffing from within sounds easier to accomplish than it is in actual practice. The ramifications of such a choice should be thoroughly considered before it is made.

Projects that can be performed entirely with resources from within are usually small, of short duration and limited scope. They can be handled within the functional organization which is the project's beneficiary. Other projects, which may involve more than one function, are still sufficiently single-function-focused to the extent little problem exists in determining where the project leadership will originate. This does not mean that the bulk of staff will come from the benefiting organization. As we discussed in the section on organizing, the project must be broken down into appropriate work elements and assigned to the best qualified individual, regardless of affiliation.

The major staffing question posed by the smaller project is whether to staff it with full-time or part-time personnel. The arguments for full-time staffing are very forceful, a primary one being the benefits obtained from the singleness of purpose and dedication to only the project. On the negative side, full-time personnel must be replaced in their current duties, creating an assimilation problem in the future and uncertaintly and insecurity for the project participants. Those who promote the part-time approach quote the adage "If you want a job done, give it to a busy man," the claim being that one can select individuals who are well-organized, efficient, and can tackle additional work without either the old job or the new job being shortchanged. This is also predicated on the fact the individual can serve both the project manager and his functional manager equally.

Experience shows that the best results in the execution of a project are obtained when the project personnel are dedicated full-time participants in the project. This assumes the full-time personnel associated with the project can concentrate on their project assignments without concern as to what they will do when the project is completed. For those with the problem of deciding whether to utilize full-time or part-time personnel I would pose the following solution. Look at the long-term picture. Assign the individual to the project on a full-time basis if a meaningful assignment, offering equal or better personal progress, can be foreseen. Care is necessary in developing the work breakdown structure to ensure the assignment is a full job. Part-time jobs may have to be combined to avoid duplication or underutilization. Individuals should be aware this policy is in force and is faithfully applied. The organizational framework should be determined before the actual selection of individuals. If the project can be handled within the functional organization, flexibility is possible. Either a project organization, reporting to the functional manager, or a matrix is possible. When the project crosses functional lines, both options are still available. Special care should be taken if the matrix is considered. If it has not been used before, a company must be willing to address the potential problems of this form or organization. Otherwise, it is not recommended. These concerns were covered in the earlier section on organization.

4.1.2   External Sources

Very few organizations have the resources to execute a project with internal resources. Experienced product development project organizations such as Proctor and Gamble still engage outside assistance in marketing and advertising although they have in-house personnel in these functions. The primary reasons for engaging outside assistance are the lack of sufficient internal resources and the desire not to add permanent staff to solve temporary needs. The proper selection of this assistance is critical to project success.

There are three major sources of contracted project services: individual consultants, service specialty contractors, and full-spectrum service contractors. Individual consultants are specialists. They are utilized as a sounding board for internally generated ideas, to provide ideas, or to suggest courses of action. They also fill a gap in expertise. On occasion they occupy temporary staff positions. Recently, it has become common for companies to engage former employees in this capacity. Specialty contractors, like the consultant, provide a specialized service, but with broader scope and capability. Firms such as market researchers, aerial surveyors, advertising agencies, and drafting job shops are typical of the hundreds of companies that offer their services to satisfy project needs. The full-spectrum contractor is just that. Short of some unusual specialty, these firms can execute an entire project with their own resources, and often do. Despite their apparent differences, selection of the right consultant, specialty, or full-service contractor is similar. It reduces to selecting qualified individuals from the contractor's organization.

The Consultant

Engaging an outside consultant is the same process that should be used in the hiring of a senior professional staff member. The first step is to develop the job specifications. The second is to determine the qualifications and experience required to perform the job. The third is to match the candidate to the job and make a selection.

Consultants should be hired to perform a specific task. What is to be done is an essential element of the job specification. This

could be: reviewing and commenting on action, assessing an opportunity, or recommending solutions to a specific problem. This
will help prevent the major problems associated with consulting.
These are the treatment of the consultant as an employee, diluting
his contribution, and overstaying his welcome.

The primary reason for engaging a consultant is to provide expertise. The emphasis in developing the qualifications for the consultant should therefore be on the experience level of the candidates. It is easier to short-list the qualified candidates if the experience can be narrowed to that applicable to the problem. Too
narrow experience, however, can result in the absence of perspective in relation to the total environment of the problems to be addressed.

Consultants can be obtained through many sources. Universities are probably the most readily available source of both expertise and experience and can also assist in expanding the search
through their association with other organizations. Trade associations also provide a ready source. Many consultants are listed in
the journals or trade magazines in their area of specialization. Most
forward-looking companies have developed a clearing house for the
many consultants and firms who make calls to solicit business. A
register is kept which includes the names and addresses of these
individuals and firms, along with a brief description of the services
or products they offer. Time is always at a premium. However,
if someone has taken the time and effort to call on you, the least
you can do is take a few minutes to find out what he has to say.
Evaluation of the caller and cataloging his wares for future recall
has proven valuable and provides a shortcut in future selection.

Screening the applicants and making a selection present the
same problems in picking a consultant as in hiring a new employee.
The information presented in response to the inquiry has been
tailored to it and presents only the most favorable data on the responder. This is a particular problem with consultants in that
aside from their experience and expertise, they are in the business of selling themselves. Separating the wheat from the chaff
can be difficult, but worth the undertaking. A well-developed
specification should help in reducing the most likely candidates to
a reasonable number. Checking the references supplied is a must
and may reveal information not provided by the consultant. Once
the potential candidates are selected, interviews should be conducted.

There are many excellent guides to proper interviewing techniques. It is a skill that should be developed by anyone with am-
bitions in management. It is an area that requires special handling.
The objectives of the personal interview should be: to confirm the
data originally provided, to answer any questions that may have

arisen as a result of the initial response, and to obtain specific information about the experience and knowledge of the candidate.
The broad range of these objectives dictates that the candidate be interviewed by more than one person. The interview should be a formalized effort, with an experienced interviewer leading the discussion, supported by those who will evaluate the functional expertise. The effort should be concentrated on determining the specific experience and expertise of the consultant that relates to the problem or project at hand. A secondary objective is to ascertain how well the consultant fits in with the project team. The project manager or a senior representative with whom the consultant may ultimately work should participate in the interview. During the interview the consultant should be encouraged to comment on the reporting relationships identified in the job specification. Problems of status are best uncovered before a relationship is started.

An old definition of a consultant is someone who borrows your watch to tell you the time and then sends you a bill. Unfortunately, this definition is sometimes correct. An early review of the relationship will avoid this and the equally frustrating problem of paying someone to confirm what you already know or want to hear. There should be some interim point at which the objectives of the relationship are evaluated and when a parting of the ways can be invoked without recourse by either party. Such an arrangement helps avoid future conflict, recrimination, and wasted time, effort, and investment by both parties. The process should be included as part of the contractual relationship.

The Specialty Contractor

Engaging an organization with a special expertise avoids the individual search and selection process, puts a wider range of experience at the company's disposal, and backs it with a corporate commitment. These advantages bring a higher cost for the services rendered, but do not necessarily guarantee satisfactory results. It is still necessary to carefully outline the job and the results that are expected and to select an individual in the organization with whom to work.

The specialized company presents an attractive alternative to finding individuals to perform a special task. Recruiting and selecting individuals with expertise in a narrow specialty has already been accomplished by the firm to be hired. The broader experience of a number of specialists provides them with an ability to analyze problems from a variety of perspectives. They also present a larger group from which to select an individual to head the effort.

The first step in selecting a company is the same as that for an individual. The job to be done, the results to be expected,

and the time frame must be accurately presented. As the effort
always involves individuals, the final selection of which firm is to
perform the work will center around an evaluation of the personnel offered. In soliciting responses to inquiries, qualifications of
key individuals are more important than the collective experience
of the company. The company may have done work for a client,
but if the individuals assigned to perform the job do not have experience with that client, they are at a distinct disadvantage.

Evaluation of the firm chosen should include comprehensive
interviews of the key individuals proposed to lead the work. A
written commitment should be obtained from the company that those
individuals selected will be made available for the job and remain
on it. All too often a company proffers its superstar in order to
obtain a contract and then it is found that the superstar has been
committed to some other project and is no longer available. Addition-
ally, key individuals are assigned and before the work is completed
they are pulled off the work to be assigned to another project.

A company responsibility to a given project provides important
additional benefits. If the problem or job to be tackled is more
complex or complicated than originally anticipated, there are additional resources which can be consulted or brought to bear. In
addition, the company usually assigns an executive sponsor to the
project. This executive is available to the client to handle personnel problems and performance problems and to circumvent confrontation between the client and the company project manager. This
executive also provides an added level of overview which resolves
issues internally before they become a problem for the client.

The Full-Service Contractor

Major projects involving the design and building of new facilities
require the scope and capacity of full-service contractors. Services may range from project definition and studies to complete design, engineering, procurement, and construction. These firms
specialize in a particular area, such as manufacturing or materials
handling, bridges and highways, or pharmaceutical plants. Others,
such as the giants of the industry, Fluor, Bechtel, and Parsons,
offer a wide range of services and employ thousands of engineers
and technicians. They are capable of the so-called "turnkey" project. That is, they can provide the complete spectrum of project
services and turn over to the client a completed facility, ready for
operation. With this variety of scope, choosing the right contractor for a given project can become a monumental task in itself.

The primary reason companies rely on outside help is that
major projects are cyclical in almost every industry. No firm can
afford the luxury of the full-time staff necessary for their

occasional occurrence or the trauma of hiring and firing which is
the other alternative. Since World War II, projects themselves have
become larger and more complex. Increased regulation and public
concern lengthen the time for completion. This has added to the
inability of companies to execute such projects with their own resources. These changes have resulted in the growth and development of many firms offering a wide range of project services.

As with consultants and specialized service companies, the objective in any search is to find the best people to do the job. It
is equally important to carefully define what the contractor is expected to do. This becomes more critical as the company decides
to perform more and more of the necessary tasks and the contractor less.

4.2   THE SELECTION PROCESS*

4.2.1   The Work Breakdown

The contractor selection process begins with a definition of what
is to be done and an estimate of the resources required to do it.
The work breakdown structure forms a basis for this definition.
The breakdown enables decisions to be made as to what the parties
will do and how they will interface. This development will uncover
inconsistencies and perhaps even change who performs certain tasks.
It will eliminate duplication, confusion, or potential interference
points. The objectives are: to ensure that all tasks are performed
by someone, that each knows his specific responsibilities, and that
the assignments make sense from the standpoint of a contractual
relationship.

4.2.2   The Resource Need

The breakdown also enables an estimate to be made of the resource
requirements and the time frame within each will be required. These
need to be determined since they will have an impact on the selection of contractors and the expected cost of the service. If the
client, with his more intimate knowledge of what he expects, does
not have an idea of the magnitude of the job, he cannot expect
the contractor to know. The objective is not a game of guess what,
but of communication. The importance of both parties knowing as
much as possible about the other's perception of the task cannot
be overstressed. If the client does not know how many people are

---

*A contractor evaluation checklist is provided in the Appendix.

needed, with what qualifications, and for how long, he should seek expert help in determining these factors.

All contractors maintain loading charts on their organizations. A typical one is shown in Figure 4.2. The basic chart provides two useful pieces of information. It shows the total number of professional personnel the contractor has on his staff and the commitment of that staff to his current work load. Professional staff includes project managers on down, but excludes secretarial, administration, and reproduction staff. It represents all personnel who would be charging their time directly to a given project and not included in overhead.

4.2.3   Prescreening

With contractors' loading charts, a first screening of potential contractors can be made. It is assumed that such charts have only been obtained from firms experienced in the type of project under consideration. The importance of determining the scope and content of the job should now be apparent. An overlay will immediately show if there is a conflict between the job requirements and the current work load of the contractor.

If the contractor's backlog of work shows that he has the capacity to handle the project, several other factors should be

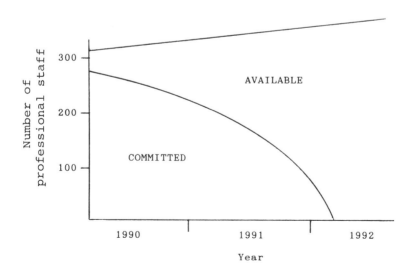

Figure 4.2   Typical contractor manpower load chart.

considered. As a general rule, it is not prudent that a single project represent too high a proportion of a contractor's work load.
Normally, this is 20% to 25%. The reason for this is that the larger
the job, the less flexibility the contractor has for controlling his
resources. Projects usually require a rapid buildup of staff. If
the project is very large, or projects currently being worked are
not finishing as expected, it may require hiring to meet the need
and prevent delays from building up. At the other end of the project, pressure is on the contractor to find work to absorb the resources being released, failing which there is pressure to retain
staff beyond the point of their real need. The ease with which
the job fits the contractor backlog provides one means to differentiate between contractors. The degree of freedom the fit provides
means that more resources are available than needed. This gives
the client a greater selection among key personnel and a higher
probability of a more rapid reaction to underestimates of requirements.

Adequacy of physical resources is only the first of several
criteria of a good prescreening effort. A determination must be
made of a contractor's qualifications, technical skills, and control
systems. Contractors should be queried as to the projects they
have done of a similar nature and their scope and responsibilities
on these projects. Descriptions of cost, schedule, and material
control systems and how they are currently being applied will provide a good indication of how projects are being executed. Checking references provides a good measure of how other clients view
the contractor and his performance. Remember, like a résumé,
everything you receive from the contractor will be intended to show
him at his best.

The foregoing methodologies can be used to develop a preliminary listing of firms that may be qualified to perform a given project or provide supplementary services. Large companies, with
multimillion-dollar capital budgets and numerous projects, maintain
regular contact with the contractors serving their industry. On
a regular basis, usually quarterly, they will solicit a report of current capacity, work in progress, and projects being tendered.
These confidential exchanges enable a reduction in the number from
which formal tenders will be solicited to those with the capacity to
handle the proposed project.

4.2.4   Contractor Development

With rare exception, contractors who provide project services are
usually selected through a competitive bidding process. Excep-

tions might include emergency projects, such as those required to restore facilities after a fire, or the exact duplication of a previous project. The objective of this process is to select the contractor who can execute the scope of work with acceptable quality, within the desired time frame, for the lowest overall cost. Attaining this objective requires an understanding that quality, performance, and lowest cost, as they relate to project services, are subjective values.

Each step of the process should be thoroughly documented and decision steps outlined. If management approval is required, having this developed, reviewed, and approved in advance of starting the process will avoid repeating parts or all of the process later.

Screening

The formal process begins with the preliminary screening. The objectives of this phase are to determine which contractors are qualified and have the capacity to perform the work and then to reduce their number. Service companies will provide a variety of data to potential clients to solicit business. Prior experience, reference to specific completed projects, and current capacity and workload can be utilized to provide a preliminary listing. For very large projects this listing may include 10 to 15 firms and for smaller projects as many as 20 firms that may be capable of doing the work. More are possible, but although they may satisfy auditors' concerns about the objectivity of the exercise, they increase the cost and duration of the evaluation. In addition, the chances of making a poor choice are increased.

The Tender Inquiry

When the preliminary listing is developed, it is time to make specific inquiries of the selected contractors. The inquiry should be structured such that the responses will be uniform to enable comparison and to match the evaluation process developed. In order to achieve this, the project should be described as completely as possible. The description should include the functional and technical scope, expected start and completion of the work, an estimate of the magnitude of the work, and any special requirements. It should also include the anticipated form of contractual relationship, such as: lump sum, reimbursed cost, and fee arrangement. This would be followed by a series of questions soliciting first a confirmation of interest in the project and then the specific items to be

evaluated. These will include: current capacity, committed workload, probable or anticipated awards, experience of a nature
similar to that of the project in question, and any special qualifications that enhance the contractor's position.

Narrowing the Choice

The responses received should be evaluated with a previously approved procedure. This is best accomplished by a team led by the
project manager. There should be some attempt at consensus, but
deference should be given to those best qualified to assess experience and technical or functional qualifications. The contractors
should be ranked in order of preference and assigned an evaluation that permits differentiation between them.

The number of contractors finally selected to submit formal
tenders is a matter of considerable debate. There are those who
suggest that all judged qualified should be allowed to tender. At
the opposite extreme are those who would pick the best and negotiate. Where the middle ground lies is a matter of opinion. Those
who would expand the list place greater emphasis on price competition and those who want to ensure the highest quality want to
narrow the choice. The thing to keep in mind is that tendering
imposes costs on all parties. The larger the number of firms competing, the lower the probability of winning the business. Firms
with higher costs are often less competitive in a pricing duel. Often these higher costs are related to a higher proportion of more
experienced employees. These are obviously the type of personnel
one would want on his project and a sound reason for not placing
too heavy an emphasis on pricing. Some firms who know a client
bases his decision primarily on price will either not expend the effort on a comprehensive proposal and submit what is called a cour-
tesy bid, or not tender at all. Unless extra quality can be shown
to have a cost benefit, the question arises as to why one should
pay a higher price for more than a job adequate to meet requirements. Narrowing the choice to the few best qualified may result
in higher costs for the same reason as cited above.

Depending on the degree of difference in the bottom end qua-
lifiers for a given job, one choice is to restrict tender requests on
very large projects to four or five, but not less than three contractors. For the small projects, this range is from five to eight.
The number assures adequate price competition while restricting
the selection to the extent one can be reasonably sure the quality
will be adequate to the need and the contractors' resources adequate to the task. Project managers get to pick the participants,

the participants get to pick the winner. Therefore, their efforts should be concentrated on assuring satisfaction working with any contractor asked to tender.

4.2.5   Contractor Selection

Once the selection of participants has been reduced, it is time to determine how the winner will be decided. Accomplishing this task requires concentration on the details and specifics for the particular project. The contractor must provide a good deal more information and the client must have a predetermined methodology for evaluation. This evaluation will include both a technical and a commercial assessment.

The Tender

At the time of formal tender, the contractor will submit an official proposal in accordance with the terms of the request. The request will include a detailed description of the project and its objectives. The tender will respond in the following specific areas:

1.   A plan of execution for the project.
2.   Resumes of key personnel with at least one alternative for each position in the top two levels of the contractor's planned organization.
3.   A staffing plan including the numbers and buildup and phasedown over time.
4.   An overall schedule, perhaps in the form of a critical path or PERT network, if the project is adequately defined.
5.   The price. This may be a lump sum. In the event of a reimbursable cost arrangement, it will include hourly rates for staff, percentages for overhead, and the fee structure.

In order to assure that a contractor fully understands the scope, content, and requirements of a project, an execution plan should be requested. The plan is a narrative description of how the contractor proposes to staff, develop, execute, and control the project. It will confirm the contractor's understanding of the resource requirements of the project. With the schedule, it will demonstrate how the proposed activities of the contractor are interrelated with those of other contractors whose services are yet to be obtained. This is certainly the case in the design of physical facilities where construction will be accomplished later, perhaps with another contractor. The plan will describe the contractors' control and reporting systems and how these are to be used to demonstrate progress to the client. If design and procurement are included in the scope of the project, an estimate of the number of engineering drawings and purchase orders may be included.

The fact should not be lost that when one hires a contractor to perform a job, one gets its people and their culture. This is particularly true with services, and in the case of project services, the choice of a firm's project manager can often make or break a project. A client should insist on the choice of who will manage the project on behalf of the contractor. The contractor, on the other hand, recognizing the importance to the project, his reputation, and indeed probable profit, should strive to offer the best he has available. The better the choice, the greater the value the client can assign to the offer. The same holds true for the project manager's key subordinates.

In order to assure that the contractor has a complete grasp of the resource needs of the project, a detailed manning chart is required. This chart is roughly the same as that provided in the prescreening inquiry, but in the tender it will include the expected requirements by discipline in specified time frames. Disciplines may include the various engineering specialties, marketing, graphic artists, aerial photographers, or other personnel, depending on the type of project. This should be coupled with similar availability charts demonstrating that the contractor has available the resources he claims are required to execute the project.

An overall schedule will demonstrate the prospective contractor's grasp of how its efforts impact and interrelate with those of others. It may be necessary for the client to provide elements of the schedule, but the contractor should integrate these into his own schedule to show dependencies and constraints. The schedule should identify the critical path, or those activities whose duration of performance determines the overall schedule. The schedule should show the lead times required for input to activities to be performed by the contractor and the delivery time of output by the contractor to dependent activities. Any conflict at these interfaces should be clearly identified and special attention called to where these are close to criticality.

Pricing of service-type contracts depends on the degree to which the content of the service can be dimensioned and the degree of risk one wants to assume or pass on. It will also depend on the kind of market one faces when going out for bids. The better defined the scope of work, the more certain the content. The risk will be lower and the probability higher that the work can be effectively done on the basis of a fixed price. Those projects where the scope is ill defined, or where alternatives are still possible, are more likely to be contracted on a reimbursable basis. Poorly defined projects present a high degree of risk of cost overruns. A client can attempt to pass off this risk by requiring a fixed price contract for the work, but should expect to pay for the contractor's acceptance of uncertainty. A soft market will often provide a high degree of owner flexibility in choosing the proper contract vehicle for a project. A tight market, on the other hand, reduces that flexibility and it may even require inducements to get the better contractors to tender.

The pricing of the tender should enable the client to determine the anticipated total cost of performing the work as defined. It will also provide a mechanism for compensation of extra work which may be accomplished during the term of the contract. This means that for lump-sum or fixed-price contracts, rates for personnel or unit prices should be provided for additional services. The client should anticipate some level of extra work and include this cost in the evaluation of the overall cost of the contract. Where rates and unit prices are provided, requirements for each should be anticipated such that the contract cost for each contractor can be developed for purposes of comparison.

Even if a project is 100% engineered, there will be areas subject to interpretation. With different goals, there will be disagreement between owner and contractor. They should be expected, there should be an adequate mechanism in the agreement to ensure they are surfaced promptly and solved quickly and equitably.

Tender Evaluation

Evaluation of contractor proposals begins at the time contracting plans are developed. When it is being decided what is to be contracted, consideration should be given to how responses are to be judged. This is particularly true if attempts are to be made to assign monetary values to noncommercial aspects of contractor proposals. Where a contractor's effort has a significant impact on the total cost of a project in comparison to its own contribution, selecting the contractor with the best technical proposal, most experienced personnel, or best execution plan will result in lower overall costs. To offset the possible higher costs of selecting such a contractor, credit for anticipated lower overall costs is assigned to its contract cost. This is an extremely subjective position and impossible of accurate assessment. If such a position is to be put forward, it should be done at an early date and the mechanism of assignment worked out, approved, and made part of the evaluation plan.

If there is not to be any commercial adjustments to contractor proposals, the evaluation of the submitted proposals has only three elements: conformance of the proposal to the tender request, technical adequacy, and expected cost. To make this evaluation effective, the commercial and noncommercial parts of the proposals should

be separated and no pricing contained in the noncommercial part.
In this way, the noncommercial aspects can be reviewed and any
questions, modifications, or corrections made before opening the
commercial proposal. Again, the objective is communication. The
firms asked to tender have been prequalified and have the where-
withal to execute the job. If there is a difference of opinion or a
misunderstanding between the parties, it is to their mutual interest
to resolve differences and remove any area of misunderstanding.
If there is an impact on the commercial terms as a result, the contractor should be allowed to submit an addendum. In the event a
difference is so significant as to be unresolvable, the contractor
should be given the opportunity to withdraw his proposal. If that
difference results from noncompliance with the terms of tender,
the client should consider rejecting the contractor as being nonre-
sponsive.

Once all of the technical issues have been conformed with, the
commercial proposals can be opened and priced out. Any exceptions to the contractual terms should then be resolved or, if the
client can accept the contractor's revised terms, priced out and
added to the anticipated cost. If exceptions cannot be resolved,
the contractor can withdraw or be declared unresponsive.

If there is to be a commercial adjustment to the technical terms,
the value and methodology should have been developed and made
part of the approved contracting plan. The most often used mechanism for such adjustments is a fixed amount as maximum and minimum to be added or subtracted from the anticipated contract cost.
Proposals are rated by a team on a numerical scale. An average
rating results in no adjustment, with above- and below-average
ratings resulting in proportional assignment of the fixed amount.

When all technical differences have been resolved, the contractual terms settled, and the expected total cost of the contract
with each contractor determined, the one with the lowest expected
cost is the winner (see the Appendix).

4.2.6   Blanket Orders

One contract mechanism that is becoming more popular is the blanket order. This is a situation in which a client has chosen a single
contractor to perform a specific set of services on a continuing
basis. The initial selection can be made on almost any basis previously discussed with a general order of magnitude of the amount
of work estimated over time. Pricing may be fixed for the short
term and subject to renegotiation. The client then merely outlines
the scope of individual projects as they are developed and issues
the contractor an order against the blanket contract.

The blanket order is a useful mechanism for a client who has a continuing need for services, but does not want to be locked into the development of internal resources to provide them. The security of such an arrangement is good from the contractor perspective in that it provides some stability from a cyclical market. Being open-ended, it poses risks for the client in that abuses are possible if the scope is not expressly defined and adequate controls are not in place. Periodic audits are a must in such arrangements.

4.2.7   A Point of Concern

Engaging consultants, specialized service companies, or broad-spectrum contractors provides the additional resources necessary to execute projects. The objective is to meld the separate parts into a functioning whole while at the same time retaining each separate identity. There is a fine line which the client, with the power of its position, must walk. That is to avoid considering the personnel of the companies it has hired as its own. The magnitude of this problem appears proportional to the number of client personnel assigned to the project. This problem can be minimized by careful attention to organizational integration, communications channels, and reporting relationships between the various organizations. As with all organizations, informal action and communications networks will develop. Such networks will function only with the support of the individuals of both parties. The problem to be avoided is the resentment of individuals who feel they work for one company and are being directed by employees of another. The client project manager is the key to avoiding this situation.

4.3   A PREFERRED APPROACH

4.3.1   Contractor Selection

The mechanisms described for engaging consultants, specialty services, and contractors are quite common, but they are not the only methods. Although I have used them and can understand and even support some of the reasons behind these procedures, I submit there is a better way. There are specific areas to which it applies. It covers those services where the output is primarily the result of mental as opposed to physical effort, that is, where the premium is placed on ideas, ingenuity, and analytical dexterity, sharpened but not constrained by experience. The scope of the work covered by such service is small in comparison to the total project. In most cases it will not exceed 10% of total project cost. It is significant to the extent that the work performed has a direct impact on the remaining cost.

A useful analogy is in the form of a question. If you had to have an operation, would you solicit competitive bids? The probable answer to this question is no. You would seek out the most qualified and experienced surgeon and may not even concern yourself with the fee. The objective is to minimize risk and maximize the chance of survival. When a company is seeking to convert a concept or idea into a reality and it can be enhanced by or requires additional help, why not get the best that is available?

The only question is what or who is best. If this question could be answered satisfactorily, it would overcome the objectives of many to direct negotiation without price competition. Who is best is a comparative evaluation and a decision of each evaluator. This is sufficient if the evaluator is responsible for the choice and the results and has the trust of his selectors. This does not mean a blank check. The contracting plan would reflect how this is to be achieved. It would be reviewed and approved at the appropriate level.

There are several benefits of direct negotiation. The negotiations and settlement can be accomplished early in the process, thereby reducing the overall duration of the project and subsequent payout. This may even be accomplished while the project concept is being developed, so the consultant or company can participate. It eliminates much of the time and cost of the competitive bidding process. How much this means is of course relative, but what it can amount to can be illustrated from experience.

In the early seventies, before the Arab oil embargo, world oil production was at maximum capacity and every barrel extracted was being sold. More in the form of an opportunity than a demonstration, a project team was set up with the charge to minimize the time to put a particular oil production facility into operation. Red tape was to be cut and cumbersome procedures streamlined. This project was executed in the period 1972 – 1973 and consumed 18 months from wildcat discovery to production. The project cost was under $5 million and had a production capacity of 180,000 barrels per day. At prices and profit levels prevailing at the time, the project had a payout of less than 3 weeks. In the mid-1980s a similar project was scheduled to take nearly 4 years. Nearly 9 months was required to simply select a design engineering contractor. Granted, other administrative niceties have crept into the execution of projects not only where this project was executed but elsewhere. It is indicative of the fact we may have lost sight of the real objective — to convert a concept or idea into a profitable reality and to begin returning the investment as rapidly as possible.

In the cited instance, construction of the project was also on a reimbursable cost basis. This enabled construction to start as soon as sufficient design drawings and materials were available. Given the profitability of the particular project and the possibility of lost market share, it was a proper action. Generally this would not be the case. It does indicate, however, that every action has a cost tradeoff. Whatever action is being planned should be subjected to a cost benefit analysis. Considering all of these possible alternatives during the planning process enhances mangement's confidence in the capabilities of the project's manager.

4.3.2   Compensation

A choice must be made of the compensation methodology for any contract. The choices are numerous and offer a variety from a fixed price to a reimbursed cost and variable fee. The objectives, however, remain the same. These are: providing compensation for costs, returning a fair profit, and offering an incentive for reduction in ultimate cost to the contractor, client, or both. Each of these objectives can be achieved in varying degrees with any type of compensation.

The fixed-price contract should be reserved for those types of efforts where the scope is firm, unlikely to change, and the risks minimal and generally within the control of the contractor. A firm scope means that the project is defined adequately. The contractor can calculate the requirements, obtain firm prices for his materials and services, and exercise a reasonable degree of freedom in execution. Risk and contingency for uncertainty should not exceed 5% of the total expected cost. This is the objective criterion. Not all situations fit and some modifications are necessary to compensate for various deficiencies.

The fixed-price contract is supposed to provide the owner with the extent of its financial exposure for the work contracted. There are many reasons an owner finds this desirable, not the least of which is the ability to fix the financing of the work and confirm the expected returns. In some cases it may even decide which projects are funded and which are not. This is usually the result of estimates as opposed to tendered prices.* The final cost is rarely what is expected, but is often within tolerable limits. What causes the variations?

---

*Clients who utilize contractor bids as a means to determine funding decisions ultimately earn a bad reputation among the better contractors. It is unfair to expect that a contractor should provide a free estimating service.

Depending on its complexity, no project scope ends exactly the
same as it starts. The more complex, the more are the changes.
Even when changes are anticipated and some control attempted
through agreed prices for labor, materials, and profit associated
with them built into the contract, implementation is never easy.
When the contract is procured in a highly competitive market, the
difficulties with changes are even greater. The reasons are apparent. Tight competition means contractors must stick close to
cost, reduce overhead, and trim contingencies for estimating errors
and risk. As soon as the work is won, the primary objective of
the contractor is to find ways to trim its costs and to stick to its
schedule. Changes are disruptive to these objectives and the owner
can expect to pay dearly for their incorporation. Even with the
protection of fixed prices for labor and materials, the owner will
still pay for normally expected idle time and higher profit margins.

A major risk on fixed-price work is reduced quality. Bid
shopping, or playing off one supplier against another to obtain
lower prices, is common among contractors. It ensures the contractor of higher margins and the owner of minimum, if not sub-
standard, quality. Diligence is called for to protect the owners'
interests. This adds to the cost of the work while not necessarily
guaranteeing results.

Not a risk, but an opportunity is lost when the benefits of
better methods or changes initiated by the contractor all accrue
to the contractor. If the results required by the contract can
still be achieved by such action, the contractor is the sole beneficiary.

Many mechanisms have been created to deal with these problems
with varying degrees of success. Fixed costs with sharing of cost
savings, penalty and bonus clauses, and combinations of these
have all been tried. Their objectives are nearly always the same:
to minimize the uncertainty of final project cost and encourage
innovation. They can be applied effectively to contracts where
the estimated final costs exceed the 5% for the well-defined project, but do not exceed 15% to 20%, the former applying to small
contracts, the latter to larger ones. Beyond these margins, the
owner can expect to pay an inordinately high price for the work
or to miss out on lower cost opportunities.

Fixed or firm pricing has little place in contracting for creative
effort. Creativity should be encouraged and the work permitted
to benefit from all the talent that can reasonably be brought to
bear. This is particularly true when a contractor is engaged to
provide project definition, design or develop a final concept, or
required to explore alternatives along with the primary objective.
The objective should be to select the best that can be found from
a reasonable number of alternatives and then manage for results.

As with any fixed-price contract, the contractor cannot be expected to do more than what is contracted. The pressure is actually on the contractor to do as little as possible and not breach the contract. No effort will be expended to benefit the client unless it will show a benefit for the contractor. Reimbursing the contractor for its costs and a reasonable profit provide the flexibility necessary to encourage creativity. Effectively monitoring such an arrangement will keep it under control. The risks associated with this type of contract put a premium on the qualifications of the client's project manager.

With the constraints of a fixed price removed, the contractor may attempt to expand his efforts. The client, under the pricing arrangement, has not only the right but the obligation to retain control of the execution process. This relationship can be effectively managed to the benefit of both the client and the contractor. If the contractor's effort is instrumental in determining other costs of the project, such as in the area of materials and construction, the benefits can be enormous.

Critics of this approach cite the uncertainty of final costs. They object to what they consider a blank check being awarded to the project's management. Granted the final costs remain uncertain, they can still be controlled. The arguments are all based on the assumption the project organization is not capable of effective management. Given appropriate selection of company personnel and effective administrative and control systems, such an assumption is without merit.

4.3.3   Motivating the Project Team

> The basic assumption in dealing with leadership style is
> that project management is more the management of the
> team than the management of the tasks.
>
> Robert J. Graham

The continuing interest in increasing the productivity of the workforce has provoked considerable research into what motivates the individual to higher levels of effort. Since the early experiments at the Bell Telephone Hawthorne plant in the 1920s and 1930s, behavioral scientists have sought to determine the factors that influence the varying levels of human activity. The secrets remain but some techniques appear to work in a variety of circumstances. Finding what works can have an influence on the success of a project and the future of the individuals involved.

4.3.4   Theories and Concepts

Frederick Taylor was one of the pioneers in the attempt to qualify the human element in the work environment [3]. The industrial revolution had created a whole new relationship of workers and their work than had previously existed in a predominantly agrarian and craft economy. Taylor's scientific management theories were based primarily on physical factors and have been mostly refuted by later work. It is not surprising that these early theories attempted to relate productivity to physical factors. It was, after all, the major element that had changed. Machinery had made possible the mass production of goods, which in turn was the foundation of the re-volution. According to Taylor, man was but an extension of the machine, and adaptation, each to the other, was the key to pro-ductivity and efficiency. Machinery was still primitive and it still required not only physical strength, but dexterity, to operate most of it. It seemed only logical that possessing physical characteristics as required by different machinery was the reason for productivity and better adaptation the answer to improving efficiency.

These theories prevailed until experiments began in the 1920s at the Hawthorne plant of the American Telephone and Telegraph Co. outside Chicago. What began as an experiment to determine how the physical environment affected productivity ended up re-futing most of the theories of the day and revolutionized behav-ioral science in the workplace. Initially, a group in this plant were isolated from the rest and light levels were varied to attempt to measure the effect of this variable on productivity. The results confounded the researchers as no matter how the light levels or other physical characteristics were varied, the productivity in-creased. They soon reached the conclusion that other factors were at work and began to look at some of the psychological factors at work. It did not take long for them to realize that the facts the group were getting special attention, allowed to set their own work pace, and similar relationships were more influential in producti-vity determination. The direction of their efforts was quickly changed, and over the next decade, the Hawthorne facility became a virtual laboratory of behavioral science and the foundation for the work which has continued unabated since.

There are two major aspects to motivation. The first are the factors influencing individual motivation. The second are those factors in the relationship of the individual and the supervisor which affect motivation. A. H. Maslow and Frederick Herzberg were influential theorists in the former and Douglas McGregor, Kurt Lewin, Rensis Likert, and Blake and Mounton, in the latter. Their work has not only provided valuable insight in explaining motivation, but has influenced many of the programs being utilized by nearly all corporations today.

A. H. Maslow theorized that humans have a hierarchy of needs, as follows:

1. Physiological needs, of which the most important is the need for food and other things necessary for survival.

2. The need for safety from danger, threat, and deprivation.
3. Social needs for association with one's fellows, for friend-
    ship and love.
4. The need for self-esteem, for self-respect, and the respect
    of one's fellows, status.
5. The need for fulfillment through development of power and
    skills and a chance to use creativity. [4]

Maslow contended that humans were motivated to satisfy each
need in turn and, as one was satisfied, would strive to fill the
next. Proponents of this theory attempt to ascertain the particular level of the individual and to influence productivity through
programs that will satisfy the next higher level of need. Empirical
research has not provided support for this theory, yet it is still
most often referenced in the literature on motivation. Support for
it is perhaps more from its intuitive strength as opposed to the
ability to confirm it formally. Have you ever wondered why a successful individual, with money and position, would spend an inor-
dinate sum to enter politics in search of a political office that pays
very little? The hierarchy theory would suggest that this individual had already filled level 4 needs and was now motivated to satisfy his level 5 needs. In this particular case, as with nearly all
political aspirants, it is a need for power.

The work for Frederick Herzberg is best known centered primarily on motivating factors. He divided these into two groups.
The first were motivators, those factors which actually motivate
people to higher levels of performance. The second, he called
"hygiene factors." These were factors such as working conditions,
pay, and supervisory attitudes. It was his theory that such factors do not motivate, but their absence can function as demotiva-
tors. He concluded, as have many since, that true motivators are
those things which enhance an individual's self-fulfillment and
self-esteem [5].

The difficulty in trying to isolate those factors which will motivate individuals is the complex relationship with the supervisor
under whom they operate. For this reason, some theorists have
concentrated their attention on supervisory attitudes, styles, and
how they are exhibited in the worker/supervisor relationship.

Much of the direction of current supervisory and managerial
training has its foundation in the work of Douglas McGregor and
his "Theory X" and "Theory Y." It was his belief that many of
the human relations problems in industry stemmed from the fact
that many managers have attitudes he called "Theory X." These
managers believe people are inherently lazy, must be prodded to
work, want security above all, and will shirk responsibility whenever they can. In contrast, McGregor suggested that managers
adopt a "Theory Y" attitude. That is, people consider work as

natural as play or rest, will not only accept but seek out responsibility, and will commit themselves to objectives when the rewards
satisfy their higher level of needs. Acceptance of these proposi-
tions was not enough in themselves, and others, such as Blake
and Mouton, have developed techniques such as the managerial
grid, to train managers to alter their attitudes in response to the
situation. The grid format was mentioned earlier and is shown in
Figure 3.2.

4.3.5   Motivating the Group

The project manager has the responsibility for motivating all the
personnel associated with the project. This includes not only his
subordinates and his company's employees, but consultants and
contractors as well. To be effective, efforts in this direction must
be active rather than reactive and premeditated rather than spon-
taneous. They must also be comprehensive as opposed to super-
ficial. Motivating a disaparate group, including those with not
only independent but diverse interests, can be a challenge and is
most assuredly hard work. The benefits to the project manager
are worth the investment. The objective is to get every individual involved in the project to feel that he is important to the project's success and is making a vital contribution which is recognized
by his peers and supervisors. Personal goals and the success of
the project should be recognized as being synergetic.

To benefit from the theories in the field of human relations,
some simple rules can be applied to motivate the project team.
These are:

1.   Maximize individual participation in the planning of pro-
     ject objectives.
2.   Permit the individual to set his own goals and establish
     how they will be achieved.
3.   Maintain consultation as a two-way street.
4.   Reward acceptable performance and correct poor performance.
5.   Praise in public, criticize in private.

The project manager will be involved with personnel at all levels.
A common denominator which has arisen from nearly all human relations studies is that motivation and participation are directly related. In addition, regardless of level, there is a high regard for
personal dignity. Every job is important and the project manager
can give recognition to this fact by involving the maximum number
of individuals in the planning process. This can be accomplished
by setting only the major parameters of the project and allowing
each successive level to develop the detailed plans to achieve the
overall goals. Individuals can then identify more closely with

the tasks at hand and have a more personal interest in their success.

Within the overall plans, each individual has his own tasks and objectives. To the extent possible, the individual should be able to establish his own goals and objectives within this overall framework. It can be the core of a "management by objectives" (MBO) program which can be used in a performance appraisal system, to be discussed later.

There are a lot of good ideas floating around if one will take the time to find them. This means getting out into the workplace and listening. It was Abraham Lincoln who said, "You learn more by listening, than you do by talking." The open door policy has several major failings. The first is it requires the talking to be done on the boss's turf, which in many instances and for some subjects is intimidating. Second, it is a constant reminder of the status difference. Third, it inhibits true, two-way exchanges.

Some call it "management by walking around"; I call it getting out where the work is [6]. Whatever you call it, there is a lot to be said for informal calls on subordinates. Your presence sends the message that you care about what people are doing. It removes the aura of your office and fosters a free exchange. Taking advantage of this and asking a few questions, probing for ideas, and really listening put the icing on the cake. Part of the success of the Japanese system rests on these very actions. They have built on the model. Supervision and management are located at the workplace. Few Japanese managers have private and isolated offices, private parking places in the company lot, or separate dining facilities. Lab coats, jackets, or uniforms are worn by all employees. These actions break down the them/us barriers, foster free communication, and enhance teamwork.

Praise and criticism are the "vitamins and minerals" of a healthy working relationship. Given in the right dosage and at the appropriate time, they can work wonders. Dispensing them is not easy. It is hard work and requires an extra effort. Knowing the best time requires prudence. Praise can be overdone and it will lose effectiveness to reinforce acceptable behavior. Criticism, on the other hand, should always be constructive and offered as soon as practical after detection of the unacceptable behavior.

Rewards and recognition satisfy higher-level needs — the need for self-esteem and the respect of one's fellows. It is important for the manager to know when praise should be between himself and the employee, or when it should be public. Praise should be a private matter, unless it is for effort that is above the average of the peer group. In that case, praise in the presence of the individual's peers will fulfill the need for respect and achieve maximum benefit.

Criticism, whenever given, should be private. It should be given as soon as practical after the unsatisfactory behavior has been detected, but in no case when the supervisor is in an emotional state. The employee should be advised of the unacceptable behavior, told what is the acceptable alternative, and offered help and advice in correcting the behavior to avoid repetition. The session should not end until it is confirmed that the employee understands what he has done incorrectly and what is expected.

A reprimand is difficult for both parties, but even more so if improper behavior is acknowledged but not addressed. It sends the wrong signals and destroys what should be a positive relationship. Most of us know when we have done something wrong, be it a pure goof or deliberate act. If you want to take the sting out of reprimand, gain points with the boss, and turn what may be a difficult situation into a plus, take the initiative. Face up to the condition, tell the boss you know what you should have done, and commit yourself not to repeat the infraction. If your boss is like most and feels uncomfortable with criticism, you will be doing him a favor he won't soon forget.

Recent studies confirm that intrinsic factors of one's job provide greater opportunities for motivation than those which are extrinsic. Griggs and Manring, for example, queried over 900 technicians to solicit what, in their opinion, motivated them [7]. The top two factors in three major categories are given below:

Job Issues
  Seeing a meaningful result from work
  Having freedom to use personal judgement and initiative
Organizational Issues
  Having opportunities to participate in choice of work
  Receiving minimum supervision and maximum latitude in
    deciding how to get things done
Career Issues
  Having individual expertise utilized and recognized
  Receiving direct, timely, and usable feedback

Every manager should recognize that the "high flier" or employee tapped for rapid advancement is already an achiever and self-motivated. These employees are few and already possessed with unusual energy. The far greater number are just average employees, need attention and encouragement, and in many cases, are unlikely to advance much further than the current job. This group provides a far greater challenge and opportunity.

Several years ago the American Management Association distributed a book to its members. Titled *Confronting Non-Promotability*, it was written by Edward Roseman and published by AMACOM in 1977 [8]. It was full of insights on how to recognize

symptoms of topping out in one's career and, more important,
.. .at to do about it. It was easy to recognize the ones that applied, but the message of the text was positive, not negative. It
was to understand that the corporate pyramid has a lot more places
at the bottom than it does at the top and only a few will make it
there. That does not mean we have failed, it only means we were
not chosen to reach the top. These are reasons why some are
chosen, others not, and in many instances they are outside the
control of the individual. The positive side is that we should recognize there are things we can do well and can enrich our jobs
so the remaining working years can be productive and enjoyable.
It's a bit tougher to do this on your own, but as a manager, you
will manage more people whose careers have stalled or topped out
than high fliers. There are things you can do to assist in recognition of this situation and to motivate people by helping enrich
their present job.

4.3.6   The Foreign Workforce

Like wines, some ideas do not travel well. Exportation of ideas
beyond national borders has not always been successful and, in
fact, has often backfired. Horror stories still aboud [9]. It is
beyond the scope of this text to advise which factors should be
considered in motivating employees in every foreign country. Caution is in order. Many concepts that work well in the United States
may meet with question when tried elsewhere. Worse, they may
meet with hostility and be counterproductive. Customs and traditions vary from country to country. In some cases, direct criticism or singling out an individual will make him ill at east and
could provoke downright hostility. For example, an individual reward to a Japanese employee for good performance may be met
with disappointment rather than pleasure. To the Japanese it means
someone else has lost. Japanese are team players and feel contribution to success is a group effort rather than an individual achievement. Recognition that the methods of any particular society are
not universal is the first step in the education process of working
in a foreign environment.

The rules of the game are different when you want to play the
game in another park. Getting to know the rules won't necessarily
assure victory but will go a long way in preventing a loss before
you get to play the game. There are many options to avoid the
pitfalls. Seeking advice from government and private agencies,
experienced in the particular country involved, is a good starting
point. Consideration of locals in key positions is another. Utilizing employees who may be nationals or former nationals of the
country can provide a dual benefit. They have knowledge of

customs and traditions and it provides an opportunity for individual development. Moving slowly, deliberately, and with preparation will minimize the pitfalls.

4.4   PERFORMANCE APPRAISAL

A control system is incomplete without a mechanism for feedback. In the control of performance that mechanism is the appraisal. Whatever system chosen, three characteristics are required: what is to be measured, how it is to be measured, and by what standard is it to be compared. Before employing any system both parties must have a similar understanding of these characteristics. The appraisal system should include all participants in the project. This includes consultants and contractors.

There are two types of appraisal system. Observant systems employ a standardized format which is completed by the supervisor. They can cover traits such as analytical ability, job knowledge, resourcefulness, and initiative. Despite the subjectivity of such a system and the fact that such appraisals are done without active input from the employee, they are by far the most common. The predominant reason for this is that they are very easy to administer and are relatively inexpensive. They are disliked by the majority of employees [10]. Interactive systems feature those characteristics required for a truly effective appraisal mechanism. The most common of these is called management by objectives, or MBO.

The project manager can implement an MBO program for his own subordinates and a modified one for the consultants and contractors under his direction. Such a program requires an investment in time and effort, but can achieve surprising results. MBO, as the name implies, is objectives oriented. Within the framework of the corporate and project objectives, the employee should establish objectives associated with his project assignment. They should be independent to the extent the employee is clearly responsible for them. The objective should vary in time duration or have intermediate measurable points to allow appraisal interchange over the entire year. The employee and supervisor should agree what constitutes an adequate performance. The system should provide for regular review and appraisal.

There is no reason why an MBO system cannot be applied to both consultants and contractors. Obviously, the system needs to be modified from that used with employees, but it will still contain the same characteristics. Where a contract is involved, objectives can be tailored to contractual commitments. Subobjectives or intermediate points can be established to permit effective

assessment over time. Within the framework of contractual obligations, agreement can be reached with the consultant or contractor
management as to what are acceptable results. The consulting
group or contractor should be encouraged to apply the system
within his own organization. Assessment should occur at regular
intervals. It provides the project manager an opportunity to formally appraise the consultants and contractors. Conversely, the
consultants and contractors should be afforded the opportunity to
provide the project manager with an appraisal of the relationship.
Where there is a performance dependency between the company and
its consultants and contractors, the MBO system can be a two-way
street, with the project taking an active part as a performer in the
system. In cases where this methodology has been used, contractor
management has been not only receptive, but highly complimentary
of the opportunity to share frank opinions of the relationship.
Communication is still the key. Retaining complaints until the end
of the job prohibits the correction of problems on which the assessment is based. Withholding compliments denies recognition when
due and, worse, the benefits that can be obtained from them.

Effective appraisal systems require direct and frequent communication. Successful communications require preparation. Richard
J. Mayer has suggested several keys to success; excerpts from
some of them are given below [11]:

1. Honesty.
2. There should be no surprises.
3. Explain the basis for your judgment.
4. Describe the performance, not the person.
5. Be concerned: If you care, it will show.

Being the client seems to instill in many project managers a
sense of infallibility. Nothing could be further from the truth.
This is increasingly the case as the client assumes a more active
role in the execution of a project. Everyone suffers from lapses
of good judgment, commits occasional errors, and doesn't always
get the job done on time. During the project, the project manager
should take stock of himself as a client. Among other things, he
should examine whether, according to Antony Jay, he or his team
have been guilty of a client's seven deadly sins [12]. These are:

1. Failure to define requirements.
2. Changing your mind and altering decisions.
3. Reacting to criticism from superiors by blaming the
   contractor.
4. Holding back real worries and criticisms.
5. Interference and second-guessing.

    6.   Blurring responsibility for others' work to take credit
        if it works and lay off blame if it doesn't.

    7.   Freeloading. Trying to get advice for nothing.

It is important not only to recognize these discretions, but to
do something about them. It is equally important to assure that
subordinates are not guilty of some of these same sins. Building
a mutually beneficial association with a consultant or contractor
transcends the current project and can prove a lasting benefit.

4.5   SUMMARY

Selection of personnel and organizations to man the project is an
important and critical element of project execution. The key individual in this process is the project manager. The project manager should be selected from the best candidates available and
carefully matched with the requirements of the project's objectives.
Talent of the kind necessary to successfully manage a project is
scarce. Do not expect the maximum, but select the best available
and support those areas where weaknesses exist. The project itself should be broken down into specific tasks and personnel assigned on the basis of their capability to handle these tasks. Personnel should be encouraged to take project assignments and as-
sured these assignments have positive impacts on their personal
development.

Choosing outside help, be it a consultant or contractor, should
be the result of a planned process. Picking this help is basically
selecting the people with whom to work and parallels the internal
selection process. This process involves: prescreening to select
those who have prior experience similar to that required for the
project, screening to develop those interested and having capacity to do the job, and a formal tender to pick the winner. There
are many mechanisms to select consultants and contractors. The
objective is to select the consultant or contractor who will make
the greatest contribution to the project.

An agglomeration of individuals must be motivated into an effective working team. The project manager is the orchestrator of
this effort and the project objectives are the focus. It has been
found that the work itself is the most effective motivator. Individuals should see the success of the project and their personal
success as linked.

The most effective appraisal system is one which includes mutually agreed objectives. Whatever system is employed, it should
provide for feedback to the individual, which includes praise, rewards, or constructive criticism as appropriate.

Special care should be taken in personnel relations with foreign employees. Customs and traditions vary. What works in one place may create havoc in another. Careful analysis of policies and procedures, with local custom and tradition in mind, should be the rule. Input from those experienced with these local factors is essential.

REFERENCES

1. Charles Martin, *Project Management*, AMACOM, New York, 1976, p. 33.
2. Robert J. Graham, *Project Management: Combining Technical and Behavioral Approaches for Effective Implementation*, Van Nostrand Reinhold Co., New York, 1985, p. 100.
3. Frederick W. Taylor, *Scientific Management*, Harper & Row, New York, 1947.
4. A. H. Maslow, *Motivation and Personality*, Harper & Row, New York, 1954.
5. Frederick Herzberg, Bernard Mausner, and Barbara Bloch Snyderman, *The Motivation to Work*, John Wiley & Sons, New York, 1959.
6. Thomas J. Peters and Robert H. Waterman Jr., *In Search of Excellence*, Harper & Row, New York, 1982.
7. Walter H. Griggs and Susan L. Manring, Increasing the effectiveness of technical professionals, *Management Rev.* pp. 62–64 (May 1986).
8. Edward Roseman, *Confronting Non-promotability*, AMACOM, New York, 1977.
9. Lennie Copeland and Lewis Griggs, Getting the best from foreign employees, *Management Rev.* pp. 19–26 (June 1986).
10. Thomas H. Stone, *Understanding Personnel Management*, The Dryden Press, New York, 1982 p. 286.
11. Richard J. Mayer, Keys to effective appraisal, *Management Rev.* pp. 60–62 (June 1980).
12. Antony Jay, Rate yourself as a client, *Harvard Business Rev.* pp. 84–92 (July-August 1977).

5
Controlling

Having carefully designed or elaborate project control systems does not of itself produce control. All of the projects over budget and schedule attest to the validity of this statement. The design of any system must start with a commitment to its objective and the involvement of all project participants. If it does not, one is left with a system that gives only the appearance of control and its implementation an exercise to perpetuate the jobs of cost engineers, planners, and schedulers.

An effective control system should ensure that what is planned to happen does happen. The initial planner must presume the actions of the many individuals who will ultimately participate in the execution of the project. Unless these later participants become involved in the plan, buy into it, and are committed to its objectives, presumptions remain just that. This does not mean that all will actually go as planned, but it offers the best chance that it will. All a control system can and should do is to determine the plan is indeed being carried out.

The junior engineers with their calculators, the draftsmen with their pencils, and the craftsmen with their tools are often the ultimate determinants of cost and schedule. Not involving them in the control process as their efforts are employed in the execution process is to protract uncertainty. That many construction contractors still do not effectively employ the tools of the critical path method (CPM) or its variants attests to the fact that those who most affect the plan have no input into it. Making it more elaborate with resource leveling options, precedence diagramming, and the like is as much a sign of frustration at the inability to get the basic concept to work as it is an improvement in the technique.

KISS, an acronym for "keep it simple, stupid," belies the benefits of doing just that. Cost, schedule, and technical control are not complicated concepts. A system to ensure that the scope, cost, and schedule objectives of a project are achieved does not have to be elaborate or complicated. Participants in the project must be intimate with it, not intimidated by it. It must be seen as a means to ensure achievement of a common goal and not a mechanism to fix blame.

> Control is the determination of progress toward objec-
> tives in accordance with the predetermined plan. Any gap
> between expectations and performance —on the part of the
> whole company or any segment of it —is most easily closed
> if it is detected before it becomes serious. Good control
> techniques provide information quickly so that action can
> be taken to correct the discrepancies. [1]
>
> Ernest Dale

Controlling is a dynamic process. The inevitability of change requires that it be a flexible process. Controlling is establishing objectives, developing controls (measurement), comparing and correcting, and providing feedback to modify the controls as required. The practice of participative management implies that the activity of controlling be delegated as low in the organization as practical.

The format of this chapter provides a general overview of the four elements of the controlling process and then application of specific methodologies to control projects.

5.0   CONTROLLING—AN OVERVIEW

> Few projects have been completed on time, within bud-
> get, and with the same staff you started with. Yours will
> be no exception. [2]
>
> Robert J. Graham

5.0.1   Objectives

Controlling a project begins with setting the goals and objectives for which the project is being executed. In order for controls to be effective, goals and objectives must be measurable. Project success or failure should not come as a surprise to its management. Neither of these should be the result of luck. Results should come about by the ability to achieve a deliberately conceived series of interim goals and objectives.

Time, money, and results, singularly or in combination, are the goals and objectives of all projects. The planning process should endeavor to establish incremental levels for each as a means to determine control of the project's execution. The ideal mechanism is the work breakdown structure. It is usually possible to assign a time, money, and result value to the individual work elements of the project. This serves the dual purpose of identification of interim objectives and assignment of responsibility to an accountable individual.

Planning economy can be achieved if overall, interim, and individual objectives are incorporated. Combining these in the plan helps to show their dependence and interrelationship. This will aid in recognizing omissions and inconsistencies. Like the pieces of a puzzle, each should have a unique identity, be small enough that the whole is recognizable in its absence, yet large enough to be missed. The balance is important in that it determines how much effort is required for control: Too large, and inertia will require a greater input to effect change. Too small, and results will not be commensurate with the effort.

5.0.2 Controls

Controlling is a process. It is a reaction to controls. Controls are passive limits within which time, expenditure, and results may occur and be considered under "control." The controls themselves do not ensure control. An example may help to distinguish the difference.

Management's objective is to control project expenditures. The project manager has prepared a budget, identifying items in detail and specifying the variation tolerances expected based on the accuracy of the estimates used. The budget has been approved, including who has authority for variations that may be needed. These are the controls. As expenditures are to be committed, they are compared to those anticipated. If they are outside the controls established, action is required. Depending on who has the authority, the expenditure may be accepted and the budget adjusted, objectives may be altered and the budget maintained. In unusual circumstances, the project itself may be dropped. This is control.

Controls are intended to determine status. They confirm the project is proceeding according to plan or is deviating. The measure of an effective control is that deviations are detectable in the first reporting period after their occurrence. They must provide a positive indication of direction, with a very high degree of reliability. Trends are important in the controlling function. Controls are designed to indicate them. The depletion of slack time in schedules or upward or downward estimates of final cost may indicate a need for action before such action is mandated by fixed control limits. The deterioration of quality may be an indication of serious deficiencies in the controls themselves. Cost and schedule estimates made by one individual or group tend to be consistent. That is, if they deviate, they will usually be high or low. As time and funds are expended, minor deviations will accumulate. Trending major factors will reveal these accumulations and signal the opportunity for early response.

Controls are the most abused element of the controlling process. To achieve project objectives, management often imposes excessive or ineffective controls. They serve only to frustrate efficient execution of the project, or tax the ingenuity of the project manager. The introduction of computers and the information explosion they have generated have compounded this excess. There is no magic formula for correction, but restraint and periodic review and assessment of the effect of established controls may be helpful.

5.0.3   Feedback

Feedback is the essence of the control system. Having established the objectives and the controls by which they will be measured, the results or status must be communicated to the decision makers. Effectiveness and efficiency depend on how far up the chain of command the communications must travel and at what level action is to be taken on the information received. The ease with which information can be generated and distributed has made the selection of information to be reported at the various decision levels and important aspect in the design of the control system. In turn, it provides opportunities to minimize the number of reporting levels by pushing decision making down the line to enhance the effectiveness of the management organization. A simple example will make it apparent that the data and reports that might be generated could be massive.

A machine operator may decide on the machine settings by measuring the variation in the output. The foreman may decide on the disposition of rejects based on the quantities on each shift. The production engineer may change the machine speed based on the number produced and the level of rejects. The production manager may decide to substitute another machine or propose the purchase of a new machine based on the unit production costs. The sales manager may request closer tolerances and lower costs as a result of customer complaints. The plant general manager may initiate a project to redesign the part and purchase a new machine based on all of the above.

It is important that feedback be considered when planning the control process. An effective method is the preparation of a process information diagram. An example is given in Figure 5.1.

The production machine has two possible outputs, acceptable or unacceptable product. Based on input from the foreman, the machine operator can determine this outcome. In many organizations this function is still being performed by a separate inspector. Enlightened

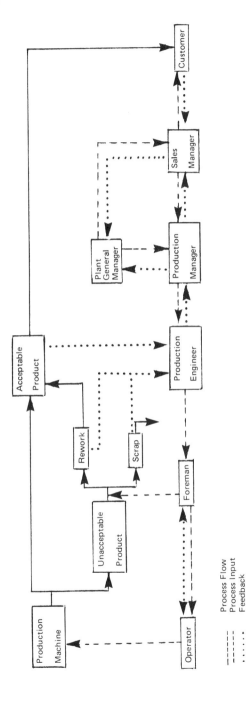

Figure 5.1   Process information diagram.

firms are finding that this function can be eliminated with no loss in quality and an increase in morale of the operating workforce.

The foreman decides, based on input from the production engineer, which of the unacceptable products can be reworked to acceptable quality and which must be scrapped. The production engineer, based on input from the production manager and feedback from the process, establishes the limits for machine speeds and tolerances — the objective being to minimize scrap and rework and to achieve the production requirements of acceptable product at minimum cost.

The plant general manager is responsible for the overall goals of the enterprise as they relate to his plant. These are established within the objectives of the business and tempered by the demands of the market. Feedback from customers assists in the development of goals. Continuous feedback maintains the critical link in the fine tuning of the operation to meet customer and corporate objectives.

It is possible with today's technology to generate and communicate an enormous amount of data. In the simplified example it is possible to collect and disseminate data regarding machine operating and downtime, speed, output, and quality. This can be by the hour, shift, and operator. How much of this is necessary and who should get it is the objective of preparing such a diagram. This determination is based on area of responsibility and the need for the information in exercising it.

All must know the overall objectives of the organization and specific goals as they relate to a particular job. The operator must know the production objectives for his job in terms of quantity and quality. If he is to judge which are acceptable and unacceptable, he must know the tolerance limits of both the output and the machine settings. The foreman must have information by which to judge which rejects can be salvaged through rework and which should be scrapped. The production engineer needs information regarding the quantities of rework, scrap, and acceptable product.

At each stage in the process an increase in the level of trust and judgment is required. Consequently, the information required to access each stage is transformed from production data to personal performance evaluation data. It is important to condense the output such that the ultimate objective is not clouded in a mass of unnecessary data.

A facet of communication that has tended to decrease as the proliferation of paper has increased is verbal communication. Management is exercised through judgment of information input. If it is accomplished only through the input in the form of reports and data collection by a staff, it becomes a staff judgment. The staff increases its importance and the manager's dependence on it by increasing the amount of paper it produces. This isolation of critical decision makers and the attendant bureaucracies which proliferate to create it have been responsible for much of the malaise of American industry. In that most bureaucratic of countries, *Glasnost* (openness) has led one Russian general director to exclaim at a Communist Party conference, "I can't stand this proliferation of paperwork. It's useless to fight the forms. You've got to kill the people producing them" [3]. Reduction of staff by many companies in the 1980s has not only cut salary costs, but brought the decision makers closer to their operations.

Utilization of a process information diagram will assist in determining the three elements of feedback: what information is to be collected, who is to get it, and what he is to do with it. As simple as it may seem, these decisions can result in a lean, informative, and responsive system, or a mountain of paper with little purpose or benefit.

5.0.4 Correction

Deviations from the set objectives are detected by the controls established and communicated by feedback. Correcting them is the final segment of the control loop. This can take many forms, from minor course correction to project cancellation. Whatever decisions are made, they should be timely and appropriate to the situation and status of the project. They should be made at the lowest level consistent with the delegation policy of the company.

Correction of the project direction can be compared to the navigation of a ship. A course has been set and heading given to the helmsman. If the ship is deviating from the heading set, the helmsman can turn the wheel to bring the ship back on heading. The navigator periodically checks the ship's position, and if the wind, currents, or geographic factors require it, he may change the heading to keep the ship on course. On occasion, the captain may dictate a change in course or a change in speed to avoid adverse weather. The ship's owner may contact the captain and change its destination to a new port of call to take on a cargo. None of these actions come as a surprise. They are anticipated and factored into the overall life of a ship at sea. Similarly, a project must anticipate that the plan will not be followed to the letter as there are factors which influence its course that are beyond accurate prediction or control. Project staff should be able to make adjustments, within their authority, in the budget and technical requirements to meet changes they encounter in design, procurement, and construction. The project manager may increase staff or emphasis in response to adverse impact on the schedule. Management may alter the project's scope in response to changing market conditions. Occasionally, as with a hurricane at sea, unusual action may be called for. All of it is in response to the feedback being received by the controls established to assure control of the situation, and not the reverse.

Examination of a complex project, which includes the elements of most projects, will show how the scope can be broken down into individual objectives and coupled with time and cost constraints. At the same time, consideration is given to the mechanisms of controls and feedback such that the seemingly disparate parts can be molded into a recognizable whole.

5.1   THE OBJECTIVES

A project has four objective categories: scope, quality, time, and budget. They are present in varying degrees of emphasis. The scope may be the functional or technical results to be achieved. Quality may be the level at which the output is expected to equal or exceed specified limits. Time involves not only the point at which the entire effort is to be completed, but should include intermediate time objectives as well. The budget delineates the funds to be spent to achieve the scope within the time allowed.

Broad project objectives may satisfy higher management needs in reviewing and approving projects. The proper planning and execution of projects, however, require that these broad objectives be the summation of detailed objectives which reach down to the individual level. Effective management begins with the individual, who must recognize his contribution and obligation to the project. This can be accomplished by setting individual scope, quality, time, and budget objectives for each project participant. These then are summarized at each level of supervision and management as the responsibility of the individual supervisor or manager.

5.1.1   The Work Breakdown Structure

A project's scope identifies the elements required to achieve the desired goal. It represents that definition of the project which management has accepted before giving its approval to proceed. Definition is restricted to only those elements required to describe the goal of the project, how it is to be achieved, how long it will take to complete, and how much it will cost. Its payout or profitability, when ranked with similarly defined projects, has successfully passed management's selection criteria and it can now be implemented. In the first phase of the implementation, the project will augment the approved scope and define a manageable project.

A complete definition of the project, its cost, and the duration of its execution are available only after the project is completed. It is one of the project manager's first jobs to conceptualize the project's details, estimate their cost, and forecast its duration, before it starts. This effort places a premium on ingenuity and experience. Like turning a picture into a puzzle, the project manager must construct the pieces, dole them out in the proper proportions, and then manage their reassembly in the proper order at the proper time. This is the objective of the work breakdown structure.

Preparation of a work breakdown is similar to unbuilding, that is, conceptualizing the elements, steps, and sequence in which the end result was created. This is one of the reasons experience plays such an important part in the effort. Having performed a similar task once, it is much easier to repeat the performance.

For illustration, assume that a company has a consumer product that has been selling well and the market is growing. It is expected that within 18 months, present facilities will be unable to meet the demand. Unless additional capacity is made available, growth will be limited and market share may be lost. Preliminary studies indicate a new plant should be built in the area of fastest market growth. It has been determined that it can be built in the time required for a cost of $3,000,000. You have just been named the project manager to implement the project.

Before you do anything else, it is necessary to understand your objective and the basis on which it was developed. This means reviewing all of the previous work which resulted in the project's definition, its timing, and its budget. For this project, it would include the process design of the plant, its physical dimensions, and the land needed. It would describe the major production machinery to be used. Without even such limited detail, a realistic cost basis would not be possible. The cost estimates will identify the elements used in their development. The time estimate will identify key activities of the project, their anticipated duration, and the cumulative total for the project.

The aforementioned elements are all expected to be present at the start. What happens if they are not? The safest assumption is something is wrong, and you must quickly fill in the gaps. That is the fundamental reason for the review. If it is not done, you have only yourself to blame when someone else is appointed to "rescue the troubled project." More about this in a later section. After satisfying yourself that all the elements are present and adequate, you can proceed with greater confidence.

The essence of a project of the size of the example may be captured with between 250 and 500 individual activities. The actual number will depend on the method of execution chosen, the contracting strategy, and how much work will be done by internal and external resources. The objective is to delineate the work into measurable activities and then to combine these into work packages. These work packages will be assigned to individuals or organizations to perform.

The most effective ways to develop a project's activities is to use the format of the logic diagram associated with the critical path method (CPM) or one of its variants. CPM is a modification of the program evaluation and review technique (PERT), which was developed by the U.S. Navy for the Polaris submarine program of the 1950s. The major difference between the two planning and scheduling systems is that PERT has a time duration for each activity that is made up of three elements: an optimistic, a pessimistic, and a most likely. CPM uses only one time duration for each activity. The logic diagram presents the activities in such a way to show the dependency of one activity on the ones which follow and the ones which precede it. Where a dependency exists and there is no real activity to show this dependency, a dummy activity with a zero time duration is used. A sample CPM is shown in Figure 5.2. It is not intended to attempt a detailed explanation here of the CPM or PERT planning and scheduling methods. There are many excellent texts available for those interested in developing a more detailed knowledge of the use and application of these techniques and their increasing variations. Suffice it to say, such a detailed knowledge is not necessary to develop a logic diagram.

The important benefit of the diagram is that it forces one to think through the project in a logical fashion. It helps to eliminate oversight and the diagramming promotes the search for alternative methods of execution which may conserve time and reduce costs. If one goes no further than this with CPM, a major portion of the benefits of the technique will have been gained.

Extreme caution must be exercised in the time estimates used for CPM. If they are averages of past experience, the probability is only 50% that you will actually complete the project on the planned date. The reason for this is the use of averages, which are the mean of all experience. There is an equal probability that the project will be completed late as early, but few have been criticized for early completion. This reinforces the need to provide a contingency in budgeting time as well as money.

As with most projects, many activities may begin at the same time and proceed in parallel. They will usually form branches which consist of groupings of activities, related to each other and easily identified. In the example project these might be: land acquisition, engineering design, and major equipment purchase. Each of these network fragments (Fragnets) will be developed to illustrate how the technique can be applied.

Land acquisition begins with a specification of requirements of the site. Among these are physical characteristics, services, and logistics. A relatively flat site is always desired, but not always available. Clearing and grading may be a significant element of cost to be considered. Services include water, sewer, fire protection, electricity, and telephone. Logistics includes road and highway access, rail siding availability if needed, proximity to commercial transportation for employees, and the location of the nearest airport. Logistics also includes the basic needs of distribution to markets and

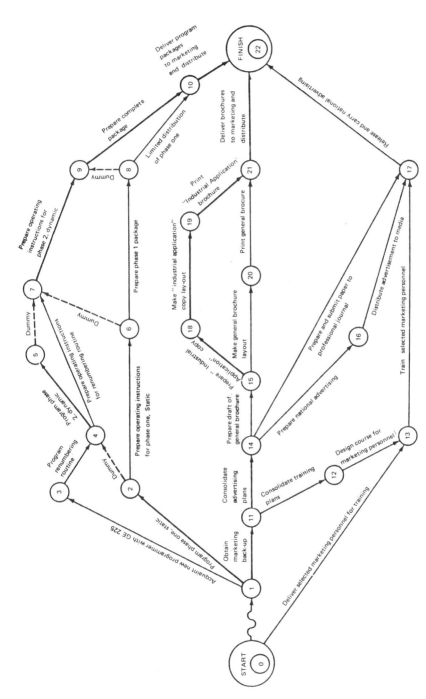

Figure 5.2 Sample critical path (CPM) logic diagram, computer program. Copyright 1983, Joseph Moder, Cecil R. Phillips, and Edward Davis, *Project Management with CPM, PERT and Precedence Diagramming*. Reprinted by permission of the publisher, Van Nostrand Reinhold, New York.

the location of customers or main distribution points to be served by
the new plant. It is apparent that if carried further, the entire
process would bog down in detail. This detail is best left to those
assigned to perform the tasks. For a project manager's purposes,
these factors may be summarized in the following activities: "site
specification, "site search," "site evaluation," and "site selection."
They can be diagrammed as shown in Figure 5.3. These activities
are serial, in that one does not start until the other is completed.
The logic diagram provides for this. If this is not logically correct,
another activity may be required reflecting the logical sequence of
the activities. For example, if alternative sites cannot be searched
simultaneously, then each alternative site could be an activity, with
evaluation also occurring at varying times. Even if the project manager knows this, he may still combine them into one activity to sim-
plify the total diagram. Care must be taken in the setting of controls for the activity to reflect what is actually going on.

In many companies, the detailed design of a new facility is beyond the capability of in-house resources. This means that an outside firm is contracted to perform the necessary design work. The
basis for this work is the process design which has been prepared
separately in the preliminary development of the project. For the
example, it is assumed that a firm is to be contracted to perform the
design, prepare the necessary construction drawings, write specifications for the construction materials requirements, procure them,
and then act as construction manager to supervise the construction.

Before the engineering design can begin, it will be necessary to
screen prospective contractors, solicit proposals, evaluate them,
select a winner, and for the selected contractor to mobilize his resources. For the project manager, these activities can be summarized in an activity called "design contractor selection."

The design phase of the project will consume a substantial portion of the total time and require interface and feedback from other
major activities. The logic of the design effort must be developed
to highlight these interfaces and feedbacks. For example, civil design, including utilities, is dependent upon site surveys, topograph-
ic drawings, and soil analysis. These, in turn, are dependent upon
selection of the site itself. Plant layouts will be prepared based on
preliminary information of the major machinery being bought, in this

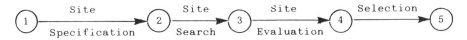

Figure 5.3 Land selection branch, project CPM, using the arrow
diagram method (ADM).

case by the company. The engineer will have a library of data
which will complement this to cover the standard items such as motor
starters, switchgear, lighting, and plumbing. Standards available in
the library will include data on the turning radii of materials handling equipment, crane clearances, and necessary personnel and
safety requirements, such as fire systems, lighting levels, ventila-
tion, and hazards such as paint spray booths, battery storage, and
so on. Unless these items are manufactured to published industry
standards, certified construction drawings are necessary. Suppliers
will not usually provide certified drawings of an item until a purchase order has been issued and frequently, not until much later.
Therefore, the layout finalization will not only depend on certified
drawings of the major equipment, but may involve contractor purchased items as well. It will not hold up progress on structural and
foundation design if the major items are attended to first.

At the start of a project it is often impossible to tell which purchased items, other than the early purchase of specially designed
equipment, may impact the design. Each design discipline must
identify the items in its area of responsibility. They will each have
a common serial sequence which will impact the overall schedule.
Typical mechanical and electrical items and their interface with the
civil and structural discipline are shown in Figure 5.4. The dummy
arrow (dashed line) indicates the dependency of the structural design completion on the receipt of certified drawing information from
the equipment vendor. Foundation design can proceed based on this
information, which includes weight, dynamic loading, and anchor bolt
layouts and dimensions.

The scheduling and assignment of responsibility for these items
are based on their dependence on each other and their respective
durations. Totally independent items will be sequenced serially,
within the allowed time frame, in as many groupings as necessary to
accommodate the staff required to oversee them. Dependent items
will be assigned to the same group, if feasible, to avoid unnecessary

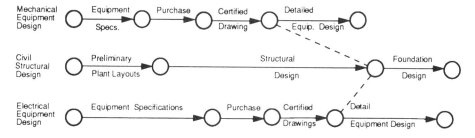

Figure 5.4   Engineering discipline equipment design and interface.

interfaces. An example would be a sump pump handling drainage from two other pumps. The specifications for this sump pump can be started based on anticipated leakage, drainage, or upset factors for the other two pumps. The final requirements cannot be determined until the details of these pumps are supplied by the selected vendor. The responsibility of the sump pump design should therefore be assigned to the same group as the two other pumps. The actual scheduling of the various activities can be dovetailed to minimize staff size and make efficient use of time.

A critical facet of the engineering design effort are the dates when certified-for-construction drawings will be available for solicitation of construction contract proposals. These are sets of drawings and specifications that will enable prospective contractors to dimension the project and offer a firm fixed price for its construction. These packages do not include all the drawings and specifications required to complete the facility, but provide enough detail to enable an experienced contractor to determine the equipment, material, and manpower necessary to complete the job and submit his firm bid.

Construction bid drawings include the plant layouts, equipment and material specifications, mechanical and electrical system details, wiring diagrams, and the standards and codes applicable to the project design and construction. Plant systems designs are produced by the engineering disciplines in parallel with the equipment designs being done by the vendors. Figure 5.5 is a modification of Figure 5.4 to show the parallel efforts of the discipline functions to produce the construction bid documents and drawings.

The first preliminary run through the project CPM should be on the basis of construction being contracted on the basis of a lump-sum fixed price. It is the most common methodology used and is nominally the least costly, given that time is available for its implementation. If it is not, alternatives must be examined. At this point it behooves the project manager to put his supervision on the alert so subsequent proposals for unusual action do not come as a surprise.

The important step when scheduling problems surface is the evaluation of alternatives. Balanced against the costs of these alternatives are the consequences of delay. The latter must be dimensioned by a risk analysis, which could include loss of sales and loss of market share. Alternatives in the scenario could include acceptance of the delay or contracting method changes. These changes may be: earlier solicitation of construction contract proposals on the basis of less complete bid requests, with acceptance of the higher prices associated with changes, or accelerated construction requiring shifting or overtime.

When the project cannot be completed on schedule within the original assumptions of execution methodology, the project manager should resist efforts to proceed without any action being taken. The

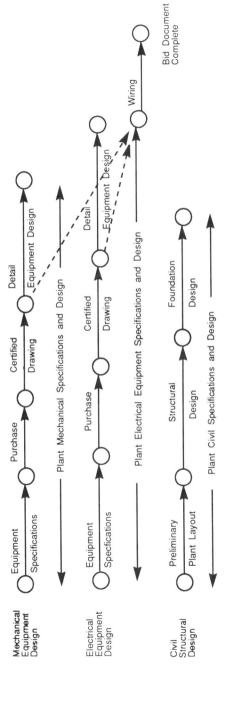

Figure 5.5   Construction bid document development.

end result of such a course is the corporate and personal cost of bringing in a project late, or backing into costly alternatives. It is then too late to use the optimum and one has to settle for a more costly solution. Aside from the company, the biggest loser in such a situation is the project manager.

The advantages of a simplified CPM for planning purposes is the ease with which alternatives can be explored and evaluated. Utilization of end points of previously dependent activities results in the longest elapsed time in the calculation of the critical path. When the need arises to shorten the duration, these dependent activities can be expanded into two or more activities whose intermediate completion will allow the subsequent activity to commence at an earlier time. This operation does not violate the logic of the network, but its increasing application results in a more complex network with more individual paths close to criticality. What is occurring is that slack is being eliminated from the network which will make it difficult later to recover any further lost time.

It can be seen how such a diagram can approach 250 to 300 activities and still be a summary. The major objectives are:

1. To confirm the ability to complete the project as conceived within the scheduled time
2. To develop and test alternatives that may result in less costly execution methodology
3. To provide a basis for allocation of work assignments
4. To establish the critical path activities and points for control of activity schedules

The work breakdown structure is then used to assign individual objectives and responsibility.

5.1.2   Individual Objectives

In the portions of the logic diagram that have been developed, groups of activities lend themselves naturally to assignment to the project manager's subordinates. They are the branches as indicated: acquiring the site for the plant, developing the engineering contract, and procurement of major equipment. These activities are independent in that they require somewhat different types of expertise and are concurrent. Except for land acquisition, they will extend for the duration of the project. Initially, they are of significant consequence and content to be pursued on a full-time basis, suggesting the assignment of three different individuals.

Organizationally it is not important whether the individuals will fit into a project organization form or a matrix. It is important that personal qualifications be considered in the selection of the individuals to perform these tasks. They will all initially be responsible to the project manager.

After selection, assignees are briefed by the project manager on the overall project objectives, the project logic diagram, the schedule, and the budget for their assigned tasks. Each will have a clearly defined task, a schedule, and budget responsibility. They should be advised of the limits of their authority and with whom they must coordinate their actions and who must concur with or approve their recommendations. Their first task, as was that of the project manager, is to confirm to themselves that the objectives assigned can be completed within the time allocated and the funds provided. They will then prepare a more detailed logic network of their own responsibility and, after review with the project manager, establish their own detailed schedule and budget within the limits of the assignment.

Each participant in the project has an obligation to confirm what has been assigned, that it can be accomplished, and to accomplish it. The work will not wait while its execution schedule and budget can be confirmed. Therefore, each participant begins by engaging in the simultaneous task of implementing his portion of the project, while confirming that it can be accomplished as approved.

5.2   CONTROL

> Particularly with technical people, the problem is not to get them to understand accounting and scheduling, but to impress on them that costs and schedules are equally important as elegant technology. [4]
>
> Charles Martin

There are three common areas of control: quality, cost, and schedule. A fourth — scope or technical control — must be assigned equal importance. Too often, emphasis on control of schedule and cost tends to mask a creeping tendency to gold-plate a project when costs are underrunning or to skimp when the reverse is true. Interplay between project participants also generates a positive or negative influence on the capacity, operating factor, and even scope of a project. It is important that the scope and technical aspects of a project be controlled independently.

5.2.1   Quality Control

Quality is more an attitude than a measure. If this is true, more effort should be expended on developing an acceptable attitude toward quality and less on the mechanics of measuring variation from specified limits. American industry is currently undergoing a transformation in just this direction. In response to the increase in competition from European and particularly Japanese products, American firms are applying statistical control methods and experimenting with quality circles and team building.

Perhaps W. Edwards Deming, more than any other single individual, has been responsible for the changing attitudes toward quality. His innovative methods and ideas were initially unappreciated by American industry. They were received enthusiastically by the Japanese and their debt has been acknowledged in the establishment of a prestigious and coveted annual prize in quality for Japanese industry and a medal from the Emperor. His ideas are simple and straightforward in their logic, which perhaps inhibited their application in American industry until recently. Too often, the simple solution to the difficult problem is overlooked. A brief capsule of the 14 principles providing the foundation of his thesis will confirm their simplicity and the proposition that quality is an attitude.

1. Create a constancy of purpose toward improvement.
2. Espouse the "new philosophy" of quality.
3. Cease dependence on inspection.
4. Award business on the basis of quality as well as price.
5. Search continuously for problems.
6. Use modern training methods.
7. Use modern methods of supervision.
8. Drive out fear.
9. Eliminate barriers between departments.
10. Eliminate numerical goals without providing the means to achieve them.
11. Eliminate work standards prescribing quotas.
12. Enhance pride in workmanship.
13. Create programs for education and training.
14. Create a top management that believe in and push for the points above. [5]

Quality control will not influence the quality of the final product if the concept of quality is not adopted and preached by the top management of the company. If the man on the production line or the craftsmen on the job site is not convinced that management cares about quality, how can he be expected to care? It has been said that in his consulting work, Deming will not proceed with an assignment unless he has a firm commitment from the company's chief executive, who is expected to be among the attendees at his first sessions. In assessing the quality control systems of any firm it is necessary to break through the fog of mere lip service and weigh the determination of management as evidenced by their attitude and the methods they have implemented to ensure quality.

For those who trust the effectiveness of inspection, I suggest a simple exercise. It involves the mere counting of the number of "g's" in the words of the following short story. With most inspections, there is only a limited time in which to check the output with the specifications and the product cannot be destroyed in the process. In this case, you have 2 minutes and cannot obliterate the "g's" in order to count them.

> While strolling through a glen, a giddy English girl tripped on a rather large, almost gigantic frog. The girl staggered but regained her footing and was about to go on when the frog began to speak and gesticulate to gain the girl's attention. "I have not always been a frog," he croaked. The frog's green coloring seemed to glow brightly as he continued, "I was once a gracious knight, a gentleman called gallant George Grenville, but was changed into this ghastly frog you now see by an ungodly, magical genie. The spell can only be broken if I gain a girl's good graces and spend a night in her garden." The agog girl was skeptical, of course. She gazed at the frog's pleading eyes and soon her giddy nature gave way to her doubts. Giggling, she decided to grant the frog's wish and took him home straightway, putting him by her garden gate. That night the girl slept grandly and sure enough, when she awoke the following morning, there alongside her garden gate was the gracious knight, George Grenville. Well, strangely enough, for a long, long time the girl's mother did not believe that story.

This exercise was what is termed 100% inspection. In most cases, only samples can be taken at random. If you answered 82, score yourself an expert, if not unique, inspector. In all the times this exercise has been given, few come up with the correct number and the range in answers is wide. The message is that even 100% inspection is an ineffective method to ensure quality.

An objective to produce quality must be backed up by appropriate methods to achieve it. For many industries, mass production techniques lend themselves to the application of statistical quality control methods. Automotive manufacturers are now insisting that their suppliers have in place such statistical control systems. Spot checks and actual supply of their vendors of control data are sufficient to ensure that these manufacturers are now receiving parts of the quality specified. The first steps are to communicate what is required and to determine how the supplier is going to achieve it.

Many items to be supplied to a project are not mass produced. In this instance, it is necessary to clearly identify what is wanted, the standard to which it is to be made, and the performance it is expected to achieve. The proper response from the vendor will be a confirmation of his understanding of the requirements and a statement identifying his quality control and testing methods by which he will verify that what he is providing is what has been requested.

As the supplier of a project to its ultimate user, you too have an obligation to provide quality. Your own attitude must be attuned to the demands of your client and your own pride in the accomplishment of your task. Whenever possible, each participant in the project should know what is expected of him and the standard to which he will be measured. In addition, that responsibility should include the ability to measure his own performance and to make corrections as necessary. These can form excellent adjuncts to the individual's management by objectives program as outlined in Chapter 3. Pride in one's work is like ownership. One must be able to identify with it. The corollary is that if something is not right, the one responsible should also be identifiable.

Inspection has its place. If statistical methods and their results are not available, actual physical inspection may be necessary. This is particularly true with the one-of-a-kind, unique output which cannot benefit from the repetitive processes of mass production. Such inspection should focus on the critical features of the particular item and start with a random selection for comparison with the requirements. The degree of inspection will depend on the criticality of the item and the extent of its conformance. If deficiencies are found, a further check of other features may be mandated.

The purchase order for any item should contain language sufficient to protect the interests of the buyer. It should stipulate that payment is contingent on conformance with the purchase requirements and these should be spelled out in appropriate detail. If payment is contingent on a final inspection, it is usually preferable to perform this inspection at the site of manufacture. If it is a complex, one-of-a-kind item, such inspections should be performed periodically to avoid the problems of critical elements being difficult to inspect after full assembly. Periodic inspection also reduces the influence of time pressure for correction and provides a basis to confirm the standards of quality expected from the vendor.

Many assembled products are made up of components, not all of which are manufactured by the vendor to whom the order is given. This places the vendor's suppliers at arm's length from the buyer. To circumvent the problems this can create, two techniques are available. The first is to require the vendor to include in his orders to subvendors the ability to inspect the subsupply. The second is to involve the vendor in the same process as you have in selecting him, that is, assessment by the vendor of the quality control of his subsupplier. If it is known that there will be extensive subsupply in the prime order, it would be well to determine how the supplier qualifies and rates his subsuppliers as to quality before selecting him.

In contracting for the supply of goods and services, as in a construction contract, care must be exercised in the area of materials supply. If specific products are to be furnished, these should be

spelled out in the tender documents. Often a buyer has a preference for specific products due to good experience with them from
prior work. Often, however, the specifications will mention a specific product in order to identify the type of performance required
and then permit "or equal." There is nothing wrong with this approach, but there is a caution. It should be required that the contractor specify the product he is supplying and if awarded the contract, he should not be able to change without the owner's permission. This prevents the abhorrent practice of "bid shopping," that
is, using the contract in hand to beat down the price of still competing suppliers or subcontractors.

5.2.2   Schedule Control

The preparation of the project manager's work breakdown structure
and CPM logic diagram provides the basis for schedule control. When
the logic has been completed, assignment of time durations will allow
determination of the activities on the so-called "critical path." These
are activities which, if not completed on time, will result in extending the project duration. All other activities are said to have
"float," or may take more time than anticipated, without affecting the
overall duration. This effort requires a computer analysis in that
calculation of the duration of all of the paths in a 250 to 300 activity
network is too cumbersome to do manually. The computer will calculate the accumulated duration of the activities in all possible paths in
the network from start to finish. This is done in both directions.
The critical activities will have a "float" of zero and all other activities will have a positive "float." The amount of this float indicates
the delay the activities on that path can incur before they themselves
become critical. It is possible to have more than one "critical path."

Once the logic confirms the project can be completed on time or
has been adjusted to accomplish this objective, controls can be established. Depending on the length of the project, the project manager should review the critical and near-critical activities to ascertain their impact and frequency. Schedule performance can be most
effectively monitored if these activities are such that at least one is
completed at regular intervals. If the project is of short duration,
say several months, this may be weekly. When the project is longer,
this may be extended to twice monthly or even monthly. It should
not extend beyond that period normally utilized by the project manager's superior to review progress. If the critical path activities do
not satisfy this requirement, selected activities can be broken down
into their components to provide additional interim critical points.
Using the previous example project, it is assumed that a first run
through the logic diagram and schedule indicated the critical path for
the first 3 months, as shown in the upper portion of Figure 5.6.
This does not have a critical activity occurring during each week of

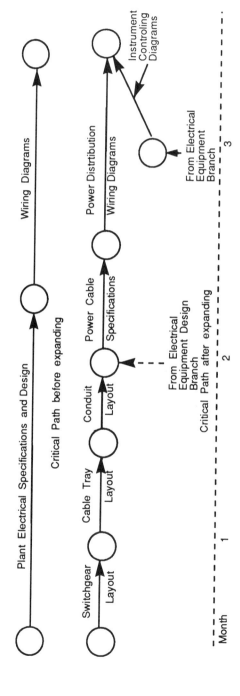

Figure 5.6   Example of expansion of critical path to enable addition of milestones.

the 2-week monitoring period. The objective can be achieved by selecting the appropriate activities and breaking them down into their more detailed components. The expanded diagram is shown in the lower portion of the figure.

This diagram does not show all of the activities involved in the plant electrical design and wiring portion of the project, only the critical ones. It also reflects the dependence of these critical activities on input from other phases of the project.

With a critical activity completing at regular intervals, the project manager has an effective tool to ascertain control of the project schedule. There are two reasons for this conclusion. The first relates to the frequency at which events must take place. Regular completion of critical activities enables the project manager to determine whether the project is ahead or behind schedule promptly enough to take appropriate action. Second, frequent and regular points, when trended, will provide an indication of tendencies of underruns or overruns of time to accumulate and foreshadow larger problems or opportunities.

The project manager is not the only one to use this technique. His subordinates should use it to control the activities of their own subordinates or contractors responsible to them for schedule performance. The project manager is responsible to higher management for overall performance of the project. Monthly reporting by the project manager to his superiors must be based on the use of this technique and be built into the formal reporting system.

Nothing is more frustrating for a manager than to find that his project is a month behind schedule, while in the previous month all was reported well. The method described will not eliminate deliberate efforts to hide bad news, but will go a long way to minimize the ability to do so. The frequency of events also tends to minimize the overall impact of any individual delay and hence encourage openness.

Before we proceed to cost control, trends of the past decades relating to scheduling should be examined. Since their development in the 1950s, PERT, CPM, and their variants have taken on the quality of a panacea. It is assumed that their use will assure on-time completion. Nothing could be further from the truth. Many contractors do not employ even the simplest of these techniques or indulge the owner because he wanted it. At the working level they ignore the program and use their own methods, like simple bar charts. Often, critical-path programs are used exclusively for reporting and then in a highly simplified summary format, reflecting less than reality. When they are used, a common occurrence is that the number of events covered is not representative of the complexity of the project. When there are too few, there is a lack of control. When there are too many, the massive output of unnecessary detail clouds the real picture. The sheer numbers of planners and schedulers engaged by contractors and clients have deluded their

managements that control is indeed present. It sends them scamper-
ing for other causes when the truth finally emerges in the form of
delays and resultant increases in cost. What has evolved in too many
cases is a planning, scheduling, and monitoring organization which,
in the parlance of Parkinson's laws, generates sufficient work within
itself to survive without any outside stimulus. The outside stimulus
in this case would be how the work is actually being planned and
executed. Utilization of CPM or PERT on complex projects involves
a commitment, sophistication, and application that is often missing.
If any of these elements is absent, the system is seriously weakened.
Many of the benefits of these techniques are gleaned in the planning
stages by those who are convinced of the benefits of the systems.
Each participant in the construction process must be made a part of
and believe in the benefits of these systems. Until this is true, the
dangers of false assumptions will prevail.

5.2.3   Cost Control

Cost control needs no elaborate definition. It is simply a system to
maintain the cost of a project within that anticipated at its inception.
The foundation of the system is the estimate of cost and the budget
for the project on which it is based. The one thing that can be
said with certainty about any estimate is that it is wrong. The final
result will not be exactly what has been predicted and will err either
plus or minus. The exact result may only be known at the very
end, when it is obviously too late to do much about it.

A project cost estimate is built from the detailed estimate of the
unit costs of all its known and anticipated elements. Some will be
known with some degree of accuracy, others will be educated guesses, and some totally unknown. The critical element in constructing
any estimate is to detail every item, including the assumptions made
for those unknowns. The importance of this becomes apparent as
the project progresses. More and more information becomes available
to refine that previously known and to illuminate the unknown. The
niche where it belongs or from which it was omitted must be identified. In other words, estimating is a continuing process, and control attempts to anticipate and affect its direction.

The budget is the funding provided to complete the project.
This consists of the estimate of costs and an allowance for uncertainty. The latter is normally called the contingency. In most control
systems, portions of the contingency are retained by various levels
of management. These are released as needed after review of commitments, expenditures, and revised estimates of final project cost.
It is the project manager's responsibility to adhere to the budget,
allocate funds to his subordinates responsible for portions of the
work, and implement the controls necessary to maintain control of
costs.

Maintaining control of costs requires the comparison of potential commitments to their budgeted allowance. They may be explicit in the budget. They may be implicit, in that they are included in a general category of expenditures, or they may have been overlooked. In the first instance, the cost that may result from the commitment is compared against its allowance in the budget. What happens with any difference will be covered later. In the second instance, a block of funds may have been allocated for the general group of purchases in which this commitment fits. This commitment identifies one of this general group and its cost will reduce the funds available for the remainder of the still unidentified items in this general group. In the third instance, funds to cover the commitment must come from the contingency established to accommodate oversight and unknowns.

Comparing potential commitments against the budget will result in three conclusions: There will either be adequate funds, excess funds, or insufficient funds. If there are adequate funds, the commitment can be made and the project can proceed. If more funds are budgeted than are required, one action is to transfer the excess to the contingency account. How this account is controlled will soon become apparent. In the event the commitment exceeds the budgeted amount, the degree of the overrun will determine the action taken.

The contingency account is the repository of funds to cover unforeseen eventualities, estimating errors and changes deemed desirable, necessary, or enhancing the return of the project. Such an account may amount to 20% to 25% of the total of a project's funds when the project is in its early stages. A table of contingency levels at various stages in project development is given in Table 7.1. Normally, contingent funds are about 10% of the total at project inception. Control of these funds are in the hands of various levels of management, who allocate their assigned portion as needs or opportunities are demonstrated. The project manager may have the authority to release as much as 25% of the total, or 2.5% of the total budget. Successive levels would have authority over equal or greater amounts. In essence, project expenditures are predetermined by the requirements of the project. What the project manager then manages are the contingency funds.

In the previous example project, items of major equipment will be purchased early in the project cycle. One of those purchases might be a tape-controlled internal grinder. Specifications have been written, quotations have been requested from suppliers, received, and evaluated. The selected supplier purchase cost is $42,000 f.o.b. factory. The budget estimate for this machine is made up of the following components:

Base Cost	$40,000
Escalation	4,000

Subtotal	44,000
Changes	2,000
Subtotal	46,000
Shipping	1,500
Installation	3,000
Commissioning	500
Subtotal	51,000
Contingency	5,100
Total	56,100

This example demonstrates the methods of handling contingency, escalation, changes, and installation or construction costs associated with major equipment. Some of these methods are also applicable to other aspects of a project.

The base cost of the equipment may be derived from previous purchase orders for similar equipment, costs from a previous project with similar equipment, supplier price lists, supplier surveys, or personnel experienced in pricing this type of equipment. Whichever source is used, the first determination is what is included in the cost used. The type of cost, such as purchase cost, installed cost, final cost, and so on, must be identified to prevent double counting. The more segregated the data used, the more accurate the estimate of similar equipment purchased and installed today.

If the cost of the equipment is a base cost, from a purchase order placed some time ago, provision will have to be made for inflation. Indices of inflation are available from many sources and for most raw and finished materials, services, and labor. The cost to purchase the item today can be estimated as the base cost of the equivalent equipment plus the inflation over the period between orders.

Most specially designed equipment costs more at the time of delivery than when initially purchased. This may result from many causes, such as the need to relocate connections, add connections and instrumentation, or modify the manufacturer's standard offer to accommodate special requirements. Prior experience is useful in determining how much this may be for the next piece of equipment purchased, but it is safe to say, much of this is pure guesswork. It should be more than zero and enough to cover the inevitable without opening the purse for gold plating.

If the equipment purchase is on an f.o.b. basis, the cost of freight must be added.

Installation costs are included in the total for this item. This is an optional choice, since they could easily have been included in the detailed estimate for the construction contract for the project. The item should not be overlooked or included twice. The reason for including it here is to highlight a common estimating technique based on highly questionable assumptions. That is the practice of "factored estimates." This practice utilizes prior experience or data of the construction costs for categories of purchased material or equipment. For example, if prior data indicate that the construction costs for the similar class of equipment are 8% of the purchase cost, then the estimated base cost is multiplied by 8% to arrive at the construction cost. This type of estimating, if used at all, should be reserved to rough preliminary estimates used to determine an order of magnitude.

Factored estimates suffer the same irrationality as the following example taken from experience with a well-known international manufacturer. Sale of standard production machinery of this manufacturer was based on published price books with variable discounts. A pump of a certain size, cast iron casing, and complete with electric driver and base was listed at $3000. Export packing of this equipment was, like a factored estimate, given in the price book as a percentage of the list price. In this case it was 5% or $150. If the customer required a stainless steel casing for this pump because of its use for a specific application, the price was $7000. Using the same procedure, the salesman applied 5% for export packing, or $350, to pack the same size pump, which in cast iron was $150.

The comparative cost for purposes of control are the budget base cost plus escalation. This amounts to:

Base Cost	$40,000
Escalation	4,000
Subtotal	$44,000

Given the supplier quotation of $42,000, there is an apparent underrun of our budget for this purchase of $2000. This is not all. The impact on the contingency allowance must be examined. If the approach to control is to be consistent, a portion of the contingency is allocable to this action. If the f.o.b. price were higher than budgeted, the contingency would have to cover the overrun. Applying our contingency factor of 10%, the amount allocable to the purchase price is $44,000 × 10%, or $4400, for a total budget of $48,400. The placement of the order for this equipment at $42,000 then results in a budget underrun of $6400. For control to be effective, something must be done with this underrun.

As a result of the placement of the equipment order for $6400 under the amount provided in the budget, the final estimated cost of the project has been reduced by a like amount. These approved, but now unneeded, funds can be freed for use to fund other projects. Obviously, this cannot be done each time a similar event occurs. An account can be established to accumulate these unneeded funds. The disposition of this account would be under the control of an appropriate level of management. It may, indeed, be necessary to tap this account for funds in the event the contingency is insufficient to cover a commitment that is significantly over the funds budgeted.

Application of the comparative analysis, which is the key to the effectiveness of the system described, takes far less effort than the preparation of an estimate. For example, the estimate for the construction of the project may involve hundreds of individual elements. If, as planned, the contract is awarded on a lump-sum basis, it is only necessary to compare the lump-sum contract price with the total of these individual elements. No useful goal can be achieved without some effort. The benefits to be gained from an effective cost control system far outweigh the effort involved in its application. The foregoing system has been found to be both effective and easy to apply and has gained the support and enthusiasm of its applicants.

Trending

Comparing cost or schedule variations to budget or time allocations alone overlooks a common characteristic of all estimates: that is, the tendency to err in a continuous direction, either plus or minus, from that originally estimated. The tendency may be very subtle and difficult to detect in individual instances. The technique of trending sequential actions serves to uncover problems before they can accumulate into irreversible conditions.

Each financial commitment results in a revised estimate of the final cost of the project, based on the difference between the commitment and the amount allocated to it in the budget. Figure 5.7 illustrates an application of the trending technique as applied to project commitments. On the X-axis are shown the cumulative commitments as they are made. On the Y-axis is shown the revised estimate of final cost based on the variations between the commitment and its estimated cost. The points are taken from the table and include the commitment for the internal grinder as the last item. The reason for using commitments is that these occur well in advance of actual expenditures and so give an earlier warning. It is assumed in projecting the resultant trend line that the remaining commitments will be as budgeted. In the example, after five commitments totaling $650,000, the final project cost will result in an underrun of $34,638. This would indicate that our project is in good shape and that our estimates tended to be on the slightly high side.

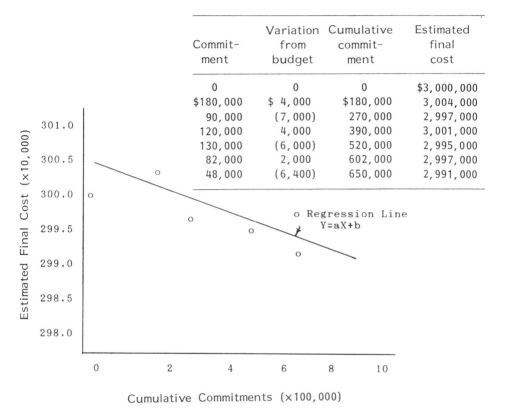

	Commit-ment	Variation from budget	Cumulative commit-ment	Estimated final cost
	0	0	0	$3,000,000
	$180,000	$ 4,000	$180,000	3,004,000
	90,000	(7,000)	270,000	2,997,000
	120,000	4,000	390,000	3,001,000
	130,000	(6,000)	520,000	2,995,000
	82,000	2,000	602,000	2,997,000
	48,000	(6,400)	650,000	2,991,000

o Regression Line
$Y = aX + b$

Estimated Final Cost (×10,000)

301.0
300.5
300.0
299.5
299.0
298.5
298.0

0   2   4   6   8   10

Cumulative Commitments (×100,000)

Figure 5.7   Sample project cost trend.

As with any projection method, there is always a margin for error. Trending provides a means to build higher and higher confidence levels in those projections as time passes. On very large projects, trends of individual classes of costs may illuminate variations that when taken together tend to cancel each other. For example, pressure vessel costs may be overestimated and mechanical equipment costs underestimated. Uncovering this fact can provide a basis for correcting the estimating base and be useful on future projects.

Given that actual expenditures lag commitments by a significant period, orders may be canceled and projects dropped if serious overruns are predicted. Cancellation charges are usually small just after orders are placed and may be more acceptable than significantly higher project costs.

A similar exercise is possible with the project schedule. As critical events are completed, days lost or gained result in a new estimate of completion. Using the same methodology, cumulative time is shown on the X-axis and the remaining estimate of time to complete, on the Y-axis. A regression analysis would indicate the time, over or under the scheduled completion, that the project may take to complete. Tailoring the critical path logic to provide additional critical activities in the early stages of the project will enhance the ability to pick up trends at the beginning of the project.

No tool is perfect and even a good system should not discourage initiative. The difference between a good manager and an outstanding one is the self-motivation and success in pursuit of better and faster ways of getting things done without waiting to be prodded. Control systems, even the most effective ones, serve mainly as caution flags and cannot, on their own, correct the potential or actual problems they reveal. A word of caution is required. The fresh look, by the newcomer or the veteran employee, which recognizes the need or desirability of change should also take into account the corporate culture. Quantum changes may not be acceptable. Change, however disguised, implies something wrong with the incumbent management. Unless that management is conditioned to change, an incremental approach may be the only way to accomplish it.

Scope and Technical Control

The scope and technical aspects of a project are the most ignored areas requiring control. When searching for the causes of project cost or schedule "creep," a likely culprit is often the subtle expansion of scope or enhancement of technical quality and performance. This occurs most often when there is a division of responsibility between the implementor and the beneficiary of the project. It is compounded when the budget is being underrun, triggering the beneficiary to embellish the scope.

At the inception of most projects, the scope and technical aspects are general and subject to interpretation. This is both a burden and an opportunity: a burden in that the implementor of the project will interpret these aspects as constraining in order to reduce project cost and execution time; an opportunity for the beneficiary to enhance the result by taking an expansive view. Even if both wear their corporate hats in their perception of the project, conflict is predictably inevitable.

The minimum project definition should contain the following elements:

1. The results expected and the input required or resources to be used
2. A description, including the size, capacity, output, volume, or horsepower of the major components
3. The operating factor and sparing philosophy to be used in sizing and selecting equipment and materials

When the project does not involve capital or physical assets, the burden of definition is on the first element. Limitation of the input resources is usually sufficient to contain expansion when coupled with the procedures that follow.

The early phases of a project is the ideal time to review the direction of implementation. It is during this time that greater definition is evolving. The original approving body, or their designates, should have the responsibility to review the project's direction. They determine whether or not that direction exceeds or falls short of the intent of the original scope.

The process is most applicable to capital projects. The most appropriate timing is on completion of the process flow diagrams, but before major commitments are made. The process flow diagram is one of the first documents produced. It incorporates the scope and defines the manner in which it is to be realized. It will contain a physical layout of the process and identify the equipment, piping sizes, flows, volumes, capacities, horsepower of drivers, and the instrumentation. It will define how the process is to work, how it is to be monitored and controlled, and what provisions exist to handle abnormalities and safety considerations.

Many firms utilize developmental reviews as the process is becoming finalized. Periodic and formalized review provides the opportunity for a broad cross-section of input and commitment by senior management and prevents major alterations occurring before their impact can be assessed by those ultimately responsible for the results. The final design basis document is prepared and signed off by the responsible members of management. It becomes the input for a more elaborate development of the design.

A methodology commonly used on major capital projects is the preparation of a "project proposal," which is an interim document developed prior to approval of project funding. It incorporates the detailed requirements implied by the design basis. An engineering contractor is often engaged to prepare it under the direction of the client company project team. It will contain the process diagram and a detailed written description of the operational and control aspects of the project. The preparation will be coordinated with other departments and with applicable standards, including those of the company. One of the objectives of this effort is to provide greater detail for estimating purposes. It is reviewed by management, modified if necessary, and approved. The estimate is then finalized and submitted for funding approval.

The project proposal process does have several shortcomings when applied in the manner described. These are readily overcome. The first is that management is not involved until the process is nearly completed. Experience indicates that conflicts most assuredly arise, and without adequate and forceful direction, they are solved by acquiescence or compromise to the detriment of the final product.

That direction must come from the project manager. The second is
the high cost of such an effort and the time required to produce it.
A solution to this problem and an assist in the former is production
of an abbreviated version of the proposal. It should be reviewed
and approved prior to development of the full proposal. Some organizations call this a "design basis scoping paper." This provides a
better definition on which to prepare the full proposal or, for shorter duration projects, an interim point at which funding approval
could be achieved.

Economy can be achieved by approaching the preparation of the
project proposal as an all-inclusive document. It provides a definition for funding approval and, if developed properly, can serve as
a bidding document for construction. This approach has saved considerable cost and time. The minimum a contractor requires for lump-
sum bidding are the dimensions of the job. A properly constructed
project proposal can contain all of this, the plot plan, the equipment
list, line sizes, instrument list, and schedule.

The project proposal serves as a control, providing a basis for
internally requested changes. The project manager's responsibility
is to execute the project as defined, and the project proposal provides the basis for this definition. Changes or modifications that
alter or amend the project should require a higher level of approval.
This eliminates capricious requests and acts as a deterrent to gold
plating.

The introduction of a "project executive" is an organizational enhancement for scope and technical control. The project executive is
a representative of the benefiting organization responsible for the
execution and budget of the project to its management. The actual
execution of the project may be assigned to another organization
within the company. They are responsible to the project executive
and look to him to approve or seek approval for actions when required. This system is very common in the oil, chemical, and pharmaceutical industries. The project executive often completes his assignment by becoming the operating manager of the facility being
built. Assignment of responsibility outside the executing organization serves to foster care and attention to both the technical and
budgetary aspects of the project.

5.3   FEEDBACK

      It seems incredible . . . that costly, scarce, and compe-
      tent human resources are still obliged to find, evaluate,
      and feed computers with up to 90 percent useless data,
      while others later have to read and control a similar amount
      of unnecessary information. [6]

Feedback is a word that evolved from process control. It is the response or status resulting from control. It has the same meaning for information as with a process response mechanism. Information feedback provides three levels of response to control. Level one is that required by those who must take direct action as a result of the information provided. Level two is that, combined with other data from other sources, required to perform other related functions. Level three is that required to provide an overview in support of broad responsibility.

Every control action requires a reaction. That reaction is the feedback or response to the control imposed. At the first level, this response is necessary to perform those functions directly attributable to project progress. If a project engineer is responsible for equipment selection, his controls are budgetary and technical. As quotations are received, it is his responsibility to obtain and review pricing and technical specifications. Unless budgetary or technical authority is exceeded, this feedback need proceed no further and an order will be placed.

When a purchase order is placed, extracts of its key elements will be passed on to others. The purchase cost and promised delivery will be communicated to the cost and scheduling engineers for incorporation into the updated schedule and cost report. It will also be passed to the finance and accounting departments, who are responsible for obtaining the funds and paying the invoice.

At an appropriate time, updated schedules, cost, and progress reports are provided to the project manager. Depending on the size and complexity of the project, these may detail all of the individual actions occurring during the report period or a summary of all events with highlights of the most significant. It is implied that no direct action is required on this type of information. It is provided as an overview and relates to the broad picture of the effectiveness of the system rather than individual events.

The information to corporate management will identify the project and state its approved budget, current estimated cost, and scheduled and estimated completion. If there are significant variations, these are explained and corrective action indicated.

The feedback methodology described makes two significant assumptions which are not always true: that the control system is working as designed and that management at all levels operates on the basis of "management by exception." The former is somewhat difficult to detect but relatively easy to correct. Detecting system deficiencies can be enhanced by simulation of anticipated problems. Fortunately, a well-designed and easily applied system is also resilient and able to cope with most eventualities. The latter needs no detection and is nearly impossible to correct. Thwarting those who want to participate in the decision process where they have already delegated responsibility or have none is a persistent problem for many project managers.

Periodic auditing of the control system is an effective way to assure it is being properly implemented. The simpler and more logical the system, the least likely that problems will develop. An audit will confirm it is working, or there are deficiencies. The audit is directed by the next level of management. Preferably, it is carried out early in the project, when most of the critical activities are being executed. It is assumed individuals assigned to the project have been thoroughly indoctrinated in the systems being employed, know their application responsibilities, and provide feedback.

Dealing with a management who wish to dabble in the responsibilities of their subordinates is a situation all will face at one time or another. The literature is full of ideas on how to manage your boss. The conclusion drawn from most is that many of their authors have never been bosses and never will be. Barring incompetence, attempts at getting back that part of your job he wants to perform can result in more undesirable side effects than the extra work involved in conforming. The system described has all the elements that anyone at any level needs to make decisions. You have at your disposal all of the facts and figures to satisfy even the most inquisitive boss. Remember, as a good subordinate, one of your jobs is to prevent the boss from making mistakes. If he insists on making your decisions, he must be convinced by the facts and figures to make the same one you would.

The third-level feedback are summaries of collected data, presented in forms that communicate their essence rather than detail. It is important to stress the substance of the information provided and how it is presented. Preparation of such summaries is part of the development and training of subordinates. It provides a supervisor the opportunity to assess the abilities of his subordinates to condense large volumes of information while capturing its meaning.

5.4   CORRECTION

Little is accomplished by the distribution of information or the provision of feedback if nothing is done with it. At a minimum this action may be passive, in that it only affects how things are done in the future. It should be required in the performance of one's job. It will provide reassurance that the project is proceeding as planned or, as this section suggests, in-course corrections are required.

Nothing will go exactly the way it is planned. This attitude prepares the project participants for the inevitable and avoids surprises. They should then be prepared to act when confronted with feedback which indicates corrections are required. The control system design should allow for correction to be invoked at the lowest possible level. A good system will detect operational variations early, while they are still modest and easily corrected.

Recognizing that no system is perfect, feedback of the correction requirements provides opportunity to refine the planning data base. The frequency, direction, and magnitude of deviation are extremely useful in revising estimates of cost and time. The system should be self-dampening. Correction, feedback, and revision of minor deviations reduce the subsequent frequency and magnitude of future deviation. The time spent correcting deviation can be better spent pursuing opportunities.

5.4.1   Budget Correction

Budget correction is more than readjustment of the budget or modification of the estimating data base to reflect current cost. The need for correction provides additional impetus to what should be a continuing effort to reduce project cost. The conservative approach is to assume that remaining activities will proceed as anticipated and the only opportunity is the one at hand.

Potential budget overruns may necessitate minor modifications in peripheral items, such as marginal instrumentation or different building materials, or major changes in process design and operating factors. Whatever is required, timing is critical. The events requiring correction are part of a continuous process and delays are likely to reduce float or add to critical schedule elements. Unless the corrections available are within the authority of the project manager, he must be prepared to:

1.   Identify the problem
2.   Dimension its impact
3.   Propose several alternative solutions, including their impact
4.   Be prepared to implement the direction chosen

The project manager should also be prepared for possible management changes in direction. One can never fully anticipate every alternative available, and in review sessions it is important to indicate which alternatives were abandoned and why. This often precludes direction to pursue alternatives that have already been considered, but rejected. Another approach is to be prepared to dimension alternative suggestions on the spot to minimize recycling. It is sometimes difficult to assemble the appropriate approving body, but getting a decision at the first sitting should be an objective. Where this is the case, there is no substitute for adequate, if not over-, preparation.

5.4.2   Schedule Correction

An advantage of a limited activity CPM logic diagram is the ability to add legitimate activities to alter the critical path. Unless the

potential delay is significant, reworking the logic should be the first
alternative in schedule maintenance. In most cases, this approach
will remove the offending activity from the critical path and restore
the schedule. Care must be exercised in this approach. If it is required too often, the result is a series of parallel paths which have
very little float, and only modest delays in a few activities result in
numerous critical activities.

If management approval of schedule alternatives is required, preparation should be similar to that for costs. Doing nothing is an
often overlooked choice. It is always an alternative, provided the
project manager is prepared to identify the consequences. Conversely, the project manager should be constantly aware of the possibilities of accelerating the project if the opportunity exists to offset its cost with project revenue. This may be possible by examining the following options for activities on the critical path:

1. Rearrangement of activities to accommodate a reduction in
    project completion time
2. Premiums to vendors for improvement in delivery dates
3. Premiums to construction contractors for additional resources,
    double shifting, or overtime on critical activities

5.5   ENGINEER, VENDOR, CONTRACTOR CONTROL

The two most effective means to control an organization
from the outside are (1) to hold its most powerful decision
maker — namely its chief executive officer — responsible for
its actions and (2) to impose clearly defined standards on
it. [7]

Henry Mintzberg

Familiarity with systems, personal rewards, and punishment make
internal implementation of controls relatively routine. Accomplishing
this objective with external organizations is not as easy. These organizations have their own systems and objectives. Interfacing so
both parties achieve their individual objectives takes ingenuity, but
is not impossible.

Nearly all contractors are capable of working a combination of
their own system and a client's system. In their relationships with
clients they must assume that clients may not have any control system, in which case they utilize their own. On the other hand, the
client may want to impose his own system, and hence they must be
prepared to adapt. Occasionally, both systems may have to be run
in parallel. Outside agencies are prepared to provide whatever the
client is prepared to pay for. If a client intends to impose his own

system, he must be prepared to pay a premium. The contract or purchase order between the parties should spell out what is expected.

5.5.1 Engineering Consultant Control

Most clients do not have internal systems to control projects. They are forced to rely on the systems of their engineering consultant. The normal relationship in this case is that the engineer functions as an agent of the client, with few client representatives, or perhaps only one, providing the interface. The problem facing the client is that of selecting the engineer with a system that will provide what it needs. It will have to satisfy internal requirements and incorporate the principles previously outlined.

At the other extreme are those organizations which execute projects as a normal course of their activities. They engage design engineers but field a shadow organization to interface with the contractor down to the second or even third level of the engineers' organization. In this case, conflicts between systems become most apparent and must be resolved before work is begun and be incorporated in the contract between the parties.

Engineering consultants maintain accurate, current, and voluminous data on the costs of material, services, and labor of construction tasks. Duplication of this effort by owners is both foolhardy and wasteful and leads users to often draw improper conclusions. There are, likewise, pitfalls in the blind use of consultant information. Experience shows that most estimates by engineers are conservative and err on the high side. This, of course, depends on the objective of the estimate. For example, if it is intended to justify a project whose approval may be in doubt, it may very well be on the low side to ensure that cost does not result in the project being abandoned. Most engineers become involved in projects that are nearly certain to be implemented, or after they have already been approved. In this case, it will look better for the engineer if the project is completed for costs below the estimate.

The data available from consultants can provide advantages, if examined in detail. In particular, the client should be aware of the additions to base costs to arrive at the final estimate. Little can be done to alter the base cost other than to be less than fully specific in what is included. Factors such as escalation, freight, contingency, and so on can inflate the base fully out of proportion to the market. Computerized estimating programs provide means to alter these add-ons. They must be examined carefully and be agreed on before accepting their addition to the base cost. The consultant's cost control system can then be employed provided the principles suggested are part of the system.

Schedules are more difficult to check, since the consultant is already likely to be prejudiced by the requirements of his client. In any event, the consultant should produce a schedule and be able to demonstrate it is achievable using his data base. This data base must contain recent project experience for the man-hours required for items such as yards of concrete, tons of structural steel, feet of cable, and so on. This is a gross check, since the actual operating schedule will eventually come from the construction contractor. The objective is a general confirmation that the project can be completed in the time required with a reasonable level of resources, working normal hours.

Compensation to the consultant is often on a reimbursable cost basis. Under this form of compensation, there is no incentive to reduce expenditures and it provides opportunity for abuse if not controlled. The two largest consultant expenditures are for personnel and materials purchased on behalf of the client. Proposals used to select the consultant should be the basis for control. These include the anticipated number of personnel to be engaged in the project, the period of their assignment, and their compensation. The contract with the consultant should stipulate that personnel will not be assigned to the project without client approval. Material procurement should be on the basis of reimbursement of purchase costs. The consultant's profit for the purchasing effort is provided by uplifts on the services necessary to perform the purchasing function. All efforts to apply a percentage to purchase costs should be strongly resisted because this compensation method encourages procurement of highest-cost items to increase compensation.

A large portion of the cost of consultants is their fee for services. This can range from a fixed fee to a percentage of billed costs and a multitude of innovative schemes in between. One of these is a sliding scale which reduces the percentage based on the number of man-hours billed. Well-defined projects lend themselves to fixed-fee arrangements. Conversely, poorly defined or development-type projects can be handled with an open arrangement.

Design consultants are in business to make a profit. Not too many years ago, relationships with consultants were similar to those enjoyed by individuals in their selection of a family doctor or dentist. If you liked him, you went to him each time you needed his services. His competence, assurance, and good service are what brought you back. If any of these were missing, you found someone else who could provide them. He in turn recognized that if he didn't give you his best, you would turn elsewhere. Competition, but primarily legal requirements, changed all this. Consultants must now bid for work on a pricing basis and this has shifted evaluation from an emphasis on quality and service to cost. It has promoted conflict between the operating and financial segments of a company over consultant selection. The result has been the concentration on price as opposed to quality in execution. In addition, the lower

cost of less experienced personnel provides an incentive to load
fixed-compensation projects with these personnel to increase profits.

Controls will not eliminate the negative influences in consultant-
client relationships but may help neutralize them. One is recognizing that labor is by far the largest element in consultant compensation. Although not universally true, it might be assumed that higher salaries attract and retain better personnel. The cost difference
between competing consultants may be the ability to offer lesser
qualified or experienced personnel. If actual compensation is used
as a basis for the proposal, the one offering the most competent and
experienced personnel would likely lose the job. This can be overcome by having the job quoted on the basis of the salary ranges
paid by the consultant for various classifications of its personnel.
This means that the range of salaries is more likely to be closer to
being equal than individual salaries. Evaluations can then be influenced in the direction of quality as opposed to price. The financial watchdogs will cringe at this subterfuge. A good response has
always been the question, if you needed brain surgery, would you
go out for bids?

Consultants are not saints. Unscrupulous management, given
the opportunity, will promise their best people at the tender stage
and then pull them back when the contract is awarded. Under the
pressure of a fixed-price contract, they will attempt to get by with
the fewest services, preferring to put their resources into reimbursable work. Given a reimbursable contract, the incentive is to load
the job with idle, less competent staff. Knowing that the pressure
is on clients to make selection on the basis of cost, the performance
on the contract in hand is not considered a factor in whether or not
they will be in competition for the next job. The consequences of
the lower performance of less competent, but cheaper staff, or the
cost of client management to offset it, cannot be accurately measured.
The situation tends to perpetuate itself, if not to grow. Assignment
of personnel from other organizations, in particular finance, may
help to gain support for a more flexible consultant selection criteria.

An often abused area of the reimbursable cost contract is that
of changes. The basis of the contract was that the effort could not
be sufficiently defined to permit a different pricing mechanism, but
this allows the possibility of uncontrolled growth. To avoid this, the
owner and contractor should agree on the basic definition of scope
and what constitutes a change. In order to keep track of the work,
changes in the scope would require an authorization and a new estimate
of the expected cost of the work. Even if there were no changes, an
estimate should be required and be given periodic review by owner
management. A confirmation by the contractor and the project manager that this represents the current anticipated costs will often
suffice to prevent uncontrolled growth in size of these contracts.

5.5.2   Vendor Control

Control of vendor supply of equipment, material, and services is
made far easier than that of consultants by the nature of the simpler
contractual relationship. Except for the area of vendor services,
the scope of the supply is well defined and the price fixed. Technical control of the supply becomes the most important factor. This
can be accomplished by specifying the standards to which the product is to be supplied and adequate inspection. As simple as this
may sound, there are still pitfalls.

Complex or long-lead purchase items usually require the production, distribution, and approval of drawings. The first are the certified layout drawings which enable the design contractor to proceed
with the plant layouts. Following these are the necessary details,
confirming compliance with the order and requiring approval by the
client or design engineer prior to the start of manufacture. Subsequent to delivery, there will be the preparation and delivery of operating and maintenance manuals. For some orders, training of client operating and maintenance personnel may be included. Control
of this process involves the schedule of all of these activities, technical approvals, and expenditures. Many buyers are now separating
the material and nonmaterial requirements of orders and requiring
separate pricing for each. The objective is to ensure that the nonmaterial requirements will be met and on time.

There are three levels of association with vendors, each of which
presents its own problems. There is the direct relationship, where
the client is the purchaser and involved in the procurement activity.
Procurement involves specifying the product, selecting the vendor,
approving drawings, etc., and inspection prior to delivery. A second is the once-removed position, where a design consultant acts on
behalf of the client. The last is where a construction contractor
supplies the equipment or material by direct purchase, to requirements contained in his contract. In this instance, the client may be
represented by his design consultant and is twice removed from the
supplier. In the case of complex machinery and equipment, the
first-level supplier may purchase components from subsuppliers, who
in turn may do the same, further separating the client from the direct supplier of any given item.

Control of supply when the client is far removed from the actual
supplier is a serious concern. This is further compounded in fixed-
price contracts where clients' attempts at control of the items supplied by contractors can result in claims of interference, with subsequent claims for extras and time extensions. The objective of any
client is to obtain quality material and equipment, even specific
items by name and model. The further removed from the actual purchase, the more difficult this becomes. Legal considerations may
make it impossible. This requires some explanation.

   Public procurement, particularly in the United States, has long
required that opportunities be open to all qualified suppliers. This
applies to both goods and services. Federal, state, and local gov-
ernments and all their agencies are required to solicit goods and
services from any source that qualifies. The intent is to discourage
discrimination in purchasing and minimize the possibilities for collusive practices. As one of the astronauts was reported to have said,
it gave him a comforting feeling, sitting in the capsule, that all of
the components around him had been supplied by the lowest bidder.
Many of these same restraints have been adopted by or imposed on
corporations. In particular, the Export Import Act and the Ribbicoff
amendment to it cover practices relating to boycotts. The particular
boycott in this case was the Arab boycott against goods supplied by
Israeli firms or firms doing business with Israel.
   The changing nature of the legal aspects of procurement would
render any specific response redundant in time. It is suggested
that procurement practices be routinely reviewed with competent legal
counsel. If specific items are deemed necessary, advice should be
obtained on how they may be bought and what the risks might be.
However, a few general techniques may be applicable.
   Duplication of existing equipment on the basis of reduced spare
parts inventory, maintenance, and training is a method by which to
justify sole source procurement. A second method is to obtain proposals from no less than three, but not more than five, of the most
qualified suppliers. If the selection can be demonstrated to be on
the basis of technical qualifications or experience of on-time delivery
and quality service, little problem will result.
   A successful method of binding suppliers of contractors and
their subsuppliers is the use of "or equal" provisions in material and
equipment specifications. This requires care and attention to the
details of procurement specifications to obtain the desired product
without exposing the client to risks of interference. The burden of
proof of equality is then on the contractor, and the client, or his
consultant, has the opportunity to judge alternate offerings.
   Areas often overlooked are installation supervision and follow-up
on startup service. This applies to complex equipment, machinery,
and systems. In some cases these services are required in support of
warranties, to ensure accurate and timely assembly or operation, and
to provide training to operations and maintenance personnel. These
services can be expensive and in some cases represent a significant
portion of the total cost. Few equipment or systems manufacturers will
provide this service on a lump-sum or fixed-price basis. As a part of
the evaluation of the cost of a given supply, it is mandatory to estimate the length of time or number of man-days this service will be required. A firm hourly rate quotation is required from the supplier
and used to evaluate the overall cost of a specific supply.

5.5.3   Construction Contractor Control

Contracting is a business practice full of risks and
controls, best managed by those sensitive to project activity and well versed in the details of project life. [8]

Robert D. Gilbreath

The majority of construction contracts are on the basis of lump-
sum firm prices or unit prices. In the rare instances when construction is on a reimbursable basis, controls similar to those employed
with design consultants are appropriate. Although the lump-sum
contractor has a contractual obligation and a financial incentive to
complete the project on schedule and at lowest cost, controls by the
client can be effective and are essential.

Schedule Control

The primary objective of schedule control is conformance to the
schedule. This is accomplished in several different ways:

1. Development of a realistic schedule which can be attained by
    a reasonable level of resources
2. Establishment of interim milestones connected to payment
    terms
3. A contractual commitment for delivery of client-supplied ma-
    terial and equipment by specified dates
4. Contractual requirements to maintain schedule performance
    supported by adequate client options to encourage
    conformance

The fixed-price agreement provides the contractor broad control
of his methods of execution. Often, this means a lower level of resources, engaged for a longer period than that anticipated by the
client. If the schedule to which the contractor is working is different from that provided in the contract, the client is faced with a
delay in project completion or undesirable legal action.

A realistic schedule is the best assurance that the one proposed
for the project will be the same as the one followed by the contractor. In developing a schedule, the client or consultant must analyze
the level of resources required to meet the target completion date.
The objective should be to set the schedule compatible with that level
of resources which will produce the lowest cost. This is exactly
what a contractor will do in bidding the job, despite the schedule
established by the client. If this completion date is not satisfactory
from an operational point of view, this must be recognized and provisions such as a bonus/penalty employed, to force earlier completions.

Interim milestones enable a more rapid response to schedule performance. Schedule slippage or acceleration can be determined at

frequent intervals and appropriate action taken as necessary. Payments keyed to completion of these interim milestones will encourage the contractor to meet them.

The client has an obligation to advise the contractor when he will deliver material and equipment he has chosen to supply. This information must be included in the tender documents to enable contractor planning and incorporated in the contract documents as a commitment by the owner. The provision offers mutual protection. The owner is protected by claims from the contractor who may alter his schedule to require an earlier delivery than is possible. The contractor has a right to expect the client to perform as contracted, or to compensate him as a result of the consequences of late delivery.

Establishment of the schedule as the essence of the contract establishes the legal basis for enforcement of contract provisions for failing to maintain schedule performance. These may include withholding of payments, requiring employment of additional resources, or cancellation of the work delayed. If the work is canceled, the offending contractor would be liable for the extra cost of having it performed by others. These are all punitive provisions, and the first objective should be a cordial relationship in which each party is encouraged to perform its obligations as contracted.

Cost Control

The agreement to complete the scope for a fixed price represents the ultimate in cost control. This is true if the scope were perfect, the specifications perfect, and the client was not permitted to change his mind. Since such an ideal situation is unlikely to occur, control of the inevitable changes is necessary. Contrary to what might be expected, many contractors do not like changes. Although they provide extra compensation with inherently higher profit margins, they are disruptive. In some cases, prices for changes are punitive, to discourage adding extra work.

A contract must incorporate the right to make changes to the work. The contractor, in turn, is entitled to fair compensation for the extra work and additional time of performance if necessary. These are agreed before the work begins. In the event agreement cannot be reached, the contract should enable the work to be mandated and its execution monitored in an attempt to reach agreement on price and schedule impact. Controversy can be avoided by providing for unit rates for extra work quantities, labor, equipment, and materials. Lump sums for the extra work may be based on agreed quantities and substituted for actual measurement of work quantities.

In recognition of the inevitability of change, evaluation of contractor tenders should include the cost of a predetermined level of changes. Evaluation of the final cost, including change costs, is

possible if the contract includes unit prices and rates. This procedure will offset the apparent advantage of low lump-sum bidders who
hope to make up the difference of an unrealistic bid with high prices
for changes. Contractors are made aware that the client intends to
employ this method of evaluation. This discourages the practice of
overpricing changes.

In addition to generalized increases in changed work prices, a
contractor will often place an inordinately high price on work quantities that he guesses may be increased or were understated. This
practice is negated by reserving the right to negotiate unit prices
prior to contract award, rejecting any and all unit prices, or, more
appropriately, assuring that the scope is adequately dimensioned.

The practice of "upsetting the bid" is another area requiring
client control. This technique is employed by contractors to price
early work higher than its true value to increase the contractor's
cash flow at the start of the job. The result is that the contractor's effort is then being financed by the client, as opposed to paying for completed work. The most effective way to discourage this
practice is to notify prospective contractors that evaluation will include an expenditure analysis. This is then carried out on the basis
of each contractor's anticipated cash flow. The cost of money or
rate of return on capital should then be applied to the difference
between the anticipated flow of funds and that reflected by the contractor's payment schedule. The difference is added to or subtracted from the lump-sum price and anticipated change order cost to
arrive at an expected final contract cost.

Caution is necessary in applying this technique. What may be
apparent from evaluation may not work out in actual practice. One
actual experience is worth nothing. The award of a substantial
lump-sum contract was made to the apparently second low bidder.
The low bidder's lump-sum advantage was offset by the stretched
payment terms offered by the next lowest bidder. To ensure this
advantage in total anticipated cost was not lost, the contract was
appropriately worded so that the contractor could not bill for any
amount over the proposed payment schedule. This payment schedule was the same as the cash flow schedule given in the contractor's
proposal and was the limit, regardless of contractor progress.

Labor Control

The arm's length status of the owner in regard to contractor labor
prevents direct control of this element of a project. No area, however, can be as important and often disruptive to an owner as that
of relations between a contractor and his labor force. Control in
this sense implies knowledge and the potential impact of good or poor
relations between the contractors selected to work on a project and
their labor force.

Regardless of the lack of any contractual position in the contractors' relations with labor, the owner often becomes a party to liability suits and a pressure point in contract negotiations between the unions and the contractor. As such, before engaging a contractor, something should be learned about the contractor's relations with labor. This includes: his history of litigation in liability actions, grievances against him and their type and the number, duration and origin of strike action. Although past history may not be repeated, it provides a good indication of what might be expected during the course of the project.

Of particular concern in unionized areas or projects that will be manned by union labor is the expiration date of current contracts. Will they all expire concurrently or are they staggered? Another question is what has been the past history of contract negotiations? Have they been amicable or have they resulted in strikes? If so, how long did the strikes last? An affirmative answer to the latter should prompt a check of the estimate to ascertain if any provision has been made to cover the impact of a probable strike during the time the project is likely to occur.

5.6 SAFETY, SANITATION, HEALTH, AND WELFARE

The increased concern for the quality of work life has placed a greater emphasis on worker safety, cleanliness of the workplace, and employee health and welfare. This concern has manifested itself in numerous pieces of legislation and governmental agencies at both the federal and state level. The Environmental Protection Agency (EPA), the Occupational Safety and Health Administration (OSHA), and the Employee Retirement Income Security Act (ERISA) are but a few. For the company and for the project, it means more than just obeying the law and filling out the necessary forms required by the many agencies. Like quality, it requires an attitude developed from really believing that things such as safety really pay.

This subject is much too complex to attempt a full treatment in a text on project management. The message for managers is clear. These are issues of deep concern; ignoring or making light of them can have severe consequences beyond the project, and they have to be tackled head on.

The first step is to develop and perpetuate an attitude of concern for the health, safety, and well-being of everyone engaged on the project. This is not only your own employees, but the employees of all the firms working on the project. It means using every opportunity to demonstrate it by words and deeds and encouraging everyone else to do likewise. On projects that are especially at risk, such as construction, contractor evaluations should place emphasis on safety records and performance. The work site should be encouraged to support safety competitions and clean-up days and pay particular attention to site sanitation, waste disposal, and personnel protection. Good efforts should be acknowledged and rewarded. Violations of rules and regulations should be promptly, adequately, and publicly dealt with.

No matter your attitude and the effort made to assure the achievement of the objectives of a good human welfare program, there are going to be problems. People at the operating level should be made aware of the potential for problems and the likely consequences. Unfortunately, there are few courses or seminars one can take and a review of cases in periodicals such as *Engineering News Record* brings into question the logic applied in the settlement of some of them. One can only be sensitive to the fact that any effort brings with it some risk and any action is likely to have potential negative consequences. The exercise of prudent judgment may not avoid all the problems, but it will eliminate most.

5.7 DAMAGE CONTROL

The term "crisis control" has been used to describe the management of catastrophes. Recent such events are the Tylenol drug tampering situation, the Bhopal gas leak in India, the Chernobyl disaster, and the Rhine River pollution resulting from a fire at a chemical warehouse. These are the extreme cases and demonstrate the need to have a plan to manage the unexpected disaster and contain its consequences. Although laudable, these situations demonstrate a need to take a more positive approach to minimize the probability that such a crisis management effort is needed. That effort might preferably be called "damage control" and should form a part of the overall control system.

Problems will develop during any project. They are a perfectly normal course of events. For the most part, they are factored into the planning effort, solved, and forgotten. Some problems defy anticipation or immediate solution and reach a stage where they have an unexpected impact. The lack of solution may actually prevent the project from operating or operational only with the assumption of some risk. In either case, additional cost in time and funds is involved. This group of problems requires focused attention.

Controlling problem resolution follows the same pattern as any planned event once its existence has become known. The problem requires definition; it has an estimated time and cost of resolution and will require feedback and action. The difference from the routine problem is the impact on the result if not resolved within the required time. There are two possible consequences of this result: the project may be delayed until the problem is resolved, or the project may be partially completed and made operational with some degree of risk. Control is focused on these consequences.

When a problem is first encountered, it becomes an event in the schedule. In addition to estimating the time and cost to solve it, a risk assessment is made as to its impact on completion. Within the categories of acceptable or unacceptable, management decides what constitutes each. A possible list is given below:

Acceptable risk	Unacceptable risk
4-week delay	Over 4-week delay
$50K cost	Over $50K cost
	Loss of life
	Pollution of any kind

Each is assigned a probability of occurrence and management determines the level of review. For example, if loss of life or pollution were involved at any probability, it requires corporate review. An expected time of solution is appended. Unlike routine project events, progress is reviewed. At each review, an updated assessment is made. The problem review level may change as circumstances change. The objective is to follow problem resolution at appropriate levels. Where risks are involved, those who must bear the consequences are given the opportunity to determine their acceptability. This system will not prevent a crisis from occuring. Its purpose is to put the selection of the tradeoffs in the hands of those who must bear the consequences of that choice.

How this might have worked in the case of the Sandoz AG Chemical Co. fire in Basel, Switzerland, is described below.

The design of any warehouse includes support systems for expected hazardous conditions. These may include temperature control, such as air conditioning, separate areas or enclosures for volatile products, and certainly sprinkler systems or at least hydrants in the event of fire. The event of fire had to be a possibility as long as flammable materials were stored. The design would therefore address the condition of what would happen to the water in that event. Estimates would have been made of the probable duration and the estimates of water volume based on system capability for the period. The absence of suitable containment for this water would have raised the natural question of possible pollution. Normal storm water run-off would also have been considered. Without containment and the possibility of ground contamination from normal spills in handling of stored materials, pollution would have been a possibility. This problem would have been raised to the highest level of company management. It would have been their decision to accept the risk of pollution or require containment.

A second line of defense was available in the event the first was not in place. The company carried insurance. Before issuing any policy the insuring agency will have examined the warehouse design and its potential contents and conducted a risk analysis of its own. This is certainly necessary to establish the probable losses and set the premiums. Pollution again would have been considered as a possibility. If measurable, the premium would have included this risk or specifically excluded it. If not, the policy would normally have excluded it as a covered risk, or at least limited the insurance liability.

This example is based on supposition. The investigations and final reports have not been reviewed. It is not even certain whether all of the facts will ever come to light or be left within the organizations who were party to it. There is no doubt that the company has put in place some mechanism to prevent this type of incident from recurrence.

5.8   SUMMARY

Controlling the results of a project is a synergistic process involving establishment of project objectives, a system of controls, arrangement of feedback of status, and correction of deviations. Objectives should be simple, measurable, and assignable to an individual who is accountable for their achievement. Controls should be economical and designed to measure deviation in time to affect correction. The feedback system should communicate status and deviation to those required to take action on it. It should cater to those who need it in the execution of related tasks and those whose responsibility it is to evaluate the performance of the participants. Correction should be a prompt response to need and action authorized at the lowest level aware of all essential facts.

Consultants, contractors, and suppliers of goods and services must be subject to control, irrespective of the contracting method. The controls complement the special obligations and requirements of these outside parties and the contractual arrangements needed to implement them. Special attention is required for reimbursable contracts, the supply of services in conjunction with that of equipment, and the peculiarities of lump-sum construction contracting. The objectives of the client are communicated to these outside parties. Agreements and understandings must be clearly stated in contract language.

The consequences of failure to resolve problems before delaying the implementation of a project or putting a plant on stream can be avoided with an adequate system of damage control. Meant to alert higher management of potentially serious problems, it formalizes problem resolution to assure considered decisions.

REFERENCES

1. Ernest Dale, *Management: Theory and Practice*, 2nd ed., New York, 1969, p. 488.
2. Robert J. Graham, *Project Management: Combining Technical and Behavioral Approaches for Effective Implementation*, Van Nostrand Reinhold, New York, 1985, p. 2.
3. Vladimir Kabaidze, in *Fortune*, p. 16 (Aug. 1, 1988).
4. Charles Martin, *Project Management*, AMACOM, New York, 1976, p. 47.
5. W. Edwards Deming, *Quality, Productivity, and Competitive Position*, Massachusetts Institute of Technology Center for Advanced Engineering Study, Cambridge, MA, 1982, pp. 16–17.
6. M. C. Grool, J. Visser, W. J. Vriethoff, and G. Wijnen, eds., *Project Management in Practice, Tools and Strategies for the 90's*, North Holland, Amsterdam, 1986, p. 62.
7. Henry Mintzberg, *The Structuring of Organizations*, Prentice-Hall, Englewood Cliffs, NJ, 1979, p. 289.
8. Robert D. Gilbreath, *Winning at Project Management*, John Wiley & Sons, New York, 1986, p. 201.

6
The Megaproject

The word megaproject entered the vocabulary to describe those projects which entailed hundreds of millions, if not billions, of dollars. Overlooked was a common thread that ran through all of these socalled projects, which is now intertwined in projects of all sizes: that is, the pervasive impact of the external environment on projects of all kinds and sizes. In this respect, projects such as hazardous waste disposal, water resources, sewage disposal, and urban development can be placed in this category.

Nuclear power plants and the Alyeska pipeline are but a few examples of the inability of earlier project management concepts to cope with the complexity faced by the modern project. No longer can a project manager be successful and survive on technical competence, a driving personality, and a modicum of interpersonal skill. He must be a diplomat, negotiator, arbitrator, and public communicator, as well as having excellent interpersonal skills. More of his time will be spent in selling the project to various constituencies than in executing it.

The complexities introduced by the megaproject place a premium on planning. The project manager must have a sweeping vision of the project's impact on both the internal and particularly the external environment. Underestimation of the existence, influence, and impact of various factions can lead to disaster. Murphy's laws are most in evidence.

The complexities and length of megaprojects might suggest that no one individual has the capability to manage one from inception to completion. Rethinking who and how responsibility and authority

are assigned may very well be in order. A planned phasing of
multiple project managers may be the answer to the problems faced
in the megaproject.

6.0 DEFINITION

The term megaproject has been with us since the 1970s. It came
about as a result of the oil embargo of 1973. The subsequent
rapid increase in the price of oil-generated, double-digit inflation
and the cost of major projects escalated in concert. What had been
a project in the millions became one in the hundreds of millions.
Those in the latter category were now in the billions. The most
famous of these was the Alyeska pipeline project in the state of
Alaska. The megaproject came to encompass all of those projects
which could be dimensioned in hundreds of millions of dollars and
more.

Defining a megaproject in terms of cost alone is misleading.
Some projects whose cost exceeds $1 billion cannot be classified
as megaprojects per se. These are indeed large projects but,
aside from their cost, can be routine from the standpoint of planning, organizing, staffing, and controlling. The distinctions of
a megaproject are the increases in interfaces and risks. These
projects involve a project manager with national or federal level
government agencies and national or international special interest
groups and expose the corporation to high levels of risk. The
manager may spend more time managing these interfaces than those
involved in design, procurement, and construction. It is this aspect which differentiates the megaproject from the norm and requires a different approach to the management of these projects.

6.1 PLANNING

> There is no widespread confidence that the giant busi-
> ness projects we are now engaged in will necessarily be
> handled much better than those of the 1970's, because
> giant-project management is still more art than science
> [1]
>
> Albert J. Kelley

A common thread in many megaprojects has been their failure
to turn out as expected. This was true with the Alyeska project.
Like almost every nuclear power plant ever built in the United
States, it was nearly four times as costly as originally anticipated.
Had these results been anticipated prior to the start of these pro-
jects, it is doubtful that many of them would have been approved

and funded. Alyeska was salvaged by the rapid rise in oil prices, increasing the return on the investment. The same cannot be said for the nuclear plants. What is it about this special breed of project that has resulted in so many turning sour?

It is reasonable to say that foresight does not have the infallible quality of hindsight. Many of the factors that cause failure can be predicted from experience or knowledge of the problems of others. How, then, could intelligent people fail to recognize the inevitable and still proceed, knowing failure was certain? One response is that the pitfalls are usually known. In eagerness to proceed, they are ignored, glossed over, or assumed to happen only to others and can be overcome. An example will support this thesis.

Lotus 1-2-3 is one of the more popular and well-known spread sheet computer programs. It is used by project controllers for accounting and budgeting of projects. Unknown to many, a program called TK!Solver was developed by Dr. Milos Konopasek. Its algorithms were called by John C. Dvorak in *PC Magazine* "a crooked bookkeeper's dream come true" [2]. This program provides a means of reverse calculation. In other words, you put in the answer and it revises the input. There is an analogy to many of these failed megaprojects. These results are preordained and the requirements to that end are generated accordingly.

This is a simple, but very plausible explanation. The aura surrounding many of these projects is so pervasive as to make it impossible for any of its proponents to consider that it may not be viable. Consciously or not, negative views are prevented from surfacing, glossed over, or countered with equally persuasive positivism. It is like an often repeated statement whose basic untruth fades with continued repetition. This is planning at its worst and is extremely difficult to circumvent or counter.

Objectivity may be the only defense against the snowballing effect of enthusiasm. In order to be successful, those who will decide the fate of a project must remain at least neutral until the case for and against are made. It is essential that both sides to the argument are developed and presented. Those who are so positive a project should proceed must also examine why it shouldn't. There is a story about the legendary Alfred P. Sloan, who chaired General Motors for over a quarter century. At a board meeting, when confronted with total unanimity and enthusiasm to pursue a certain course, he cautioned against premature action. Time did not influence the decision. He suggested the members consider why the action should not be pursued and that they raise the issue again at the next meeting. By the next meeting, unanimity had given way to skepticism and the proposal was dropped [3]. This type of reflective thinking may not have stopped all of the overrun or delayed megaprojects, but would have resulted in their being more accurately dimensioned.

A balanced approach to planning these projects is also needed. There are enough horror stories to provide sufficient cause to temper cost and time estimates. The planning should include a program evaluation and review technique (PERT) approach to cost, scheduling, and risk analysis. This technique provides for an optimistic, and most likely estimate. This is not to suggest that PERT itself be used as the scheduling technique, but it must be remembered that utilization of only estimated average activity durations to develop a critical path leads to problems. This was covered more fully in Chapter 5. A special effort is needed to give full consideration to all of the factors that can go wrong and sufficient contingency to allow for those which are not readily apparent. As one of Murphy's laws states, "Beware the hidden flaw." As any project needs a companion, the megaproject needs a conscience.

The scope, magnitude, and risks associated with megaprojects mandate the development of a strategic plan for their analysis, evaluation, and implementation. Public impact cannot be overlooked or overstated. Therefore, particular attention is necessary to those elements which in most cases are missing or less critical in the normal project. These include:

1. Local support
2. Special interest groups
3. Environmental impact

Before I comment on any of the above elements, a word of caution is necessary. Reliance on secrecy, intrigue, or attempts at denial are doomed to failure. Before the project gets too far along, many people will be aware of it. As the number of knowledgeable or even suspicious people increases, the risk of the project becoming common knowledge also grows. Good intentions and all of the sound explanations notwithstanding, secrecy will be taken as a reason to hide something. Suspicion will have been created and will cast a pall over project communications throughout its life.

6.1.1   Local Support

With few exceptions, megaprojects cause significant change in the communities in which they are implemented. Managing this change should occupy a significant portion of the early effort. The success, or failure, of this effort will establish the climate to which the project will be subject during its lifetime. Recognition must be given that change in itself is perceived as threatening. It will be resisted. People much prefer the status quo. Overcoming this resistance and quieting the fears requires talent, ingenuity, and, most of all, patience, Once again, planning comes to the rescue.

Knowing the opposition is important, but only half the equation of success. One must also know who are the potential supporters and how to marshal them to convert the undecided and neutralize the opponents. Positive efforts must be made to influence the neutrals since this is usually the largest group and the one which usually determines the outcome. This knowledge comes from in-depth research of the local media to ascertain who the leaders are and where they stand on issues affected by the project, as well as which local organizations, such as fraternal organizations, environmental groups, and labor unions, are active and what their reaction to the project might be.

Planning an approach to the local community must include both those who might support and those who might oppose the project. It must also give recognition that there are always negatives, at least from the perspective of some individuals. To delude yourself that this is not so foreshadows certain failure. The project's positive aspects need to be highlighted and emphasized. The negative aspects need to be recognized and openly discussed. The result that should be targeted is that the audience, of whatever persuasion, is convinced that the positive outweighs the negative.

The plan to gain community support of the project should cover all media and involve the highest levels of management. It must be tailored to reach the whole community, not just key supporters or potential opponents. No age group, minority, or special interest group should be overlooked. TV, newspapers, radio and print media for distribution all provide effective means of communication to a broad spectrum of the population.

Well-rehearsed and well-prepared senior management can make presentations tailored to specific audiences. This is important when the audience is likely to oppose to the project. Speakers must be prepared to answer tough and hostile questions. If answers are not readily available, don't ad lib. Commit to providing them as soon as possible and follow up to ensure they are.

If the state or community in which the project is planned has a business development counsel, this is a good place to dry-run plans. Such groups will receive news of a major project being planned in their area with a positive and helpful attitude. They may also help to confirm your list of potential supporters or detractors. Many states and localities offer special incentives for investment, such as free or low-priced land, utilities, tax concessions, employee training subsidies, and other benefits. These organizations are also good sources of current information on the prospective location, such as demographics, employment levels, public services, and education facilities.

Setbacks in any plan are to be expected. When public acceptance plays a key part in the progress of the effort, optimism is
tempered with the probability that some, perhaps many, will not
share enthusiasm for the project. The preparation of a solid foundation for the project and a positive attitude toward its future operation will take a substantial investment in time and money. Unfortunately, many bad examples can be cited to convince management of the need for sound planning. Solid waste, sewerage, and
nuclear power plant projects provide a bevy from which to choose.

Several issues are critical to the development of local attitudes
toward a major development project:

1. Changes in the economy
2. Impact on Services
3. Changes in the environment

All of these are two-edged swords. Each contains elements that
can be argued positively and negatively. There are those who
would welcome the influx of new customers, the promise of more
jobs, and a revitalizing of the community. On the other hand,
some see it as an increased demand giving rise to higher prices
and shortages in basic needs. There is little doubt that a major
project, bringing more jobs, will decrease unemployment and cut
welfare costs. An increase in the tax base will help fund needed
services, yet some will see these as strained, such as schools,
sanitation, and transportation. Planning the message to the community must address these issues and include positive action to
support the apparent benefits and overcome the negative percep-
tions.

Almost every megaproject involves a major construction effort
with its attendant temporary demands. In a large metropolitan
area, these demands can be absorbed and met with existing resources. In a smaller community, these demands can be devastat-
ing. Demands for goods and services will outstrip supply, causing
shortages and inflation of prices. Of even greater concern is the
influx of perhaps hundreds of construction workers, increasing
pressure on housing and services. Planning must consider sharing by the community of the benefits resulting from the temporary
prosperity of a large construction effort. Working with local community leaders and groups should focus on minimizing the impact
while preventing polarization.

6.1.2   Special Interest Groups

The temporary, complex and often loose nature of the
relationship and authority patterns involved in project
work, combined with the number of different departments

and companies involved in any one project, whose objectives and management style may differ, leads to human
behavior problems and a tendency for conflict between
groups and individuals. [4]

                                                                    F. L. Harrison

During the planning stage, special interest groups are identified and strategies developed on how to communicate with them.
Each area will have its own cadre of special interest groups, but
nearly all will include: labor unions, small business, environmentalists of several persuasions, political parties, and religious de-
nominations. Many have national and international affiliations, and
it is expected that their interests will have many pressure points
far afield from just the location you may be considering. Support
for a project must be addressed on a broad basis. Is is important
to maintain a global approach and not allow the project to become
an issue between various factions. Communication of the benefits
and explanation of some of the concerns about the project are structured in such a way as not to appear directed at any group.

The Republicans and Democrats, the employed and the unem-
ployed, business and labor already have sufficient reason to hold
opposing views and objectives. Remaining neutral in this environment may be difficult, but not impossible. Presentations and explanations of the project and its benefits require careful review
to ensure that something positive can be noted by all. If there
are deep divisions between groups, they will seek to find anything
that can be used to widen the gulf. Additional cannon fodder must
not knowingly be provided.

Particular care must be taken in open public forums to avoid
amplifying local differences. Preparations are essential for questions that may give rise to the expression of strongly held opposing views. Avid and vocal supporters of special interests are the
most likely audience at any public gathering. Ducking tough issues will only increase suspicion. The objectives are to take the
lead and be proactive. The speaker must maintain the momentum
and retain control of both the issues and the situation. If the
situation gets out of control, a course of action is retreat and re-
trenchment. This may be accomplished by agreeing to meet with
the most vocal objectors separately. These meetings must be formal, well documented, and reported. Such action takes away the
normal audience to whom many opponents play. The reporting of
results undercuts the opportunity for distortion.

Several methods are available to help presenters maintain control of the situation when it is expected that the reception might
be anything but cordial. Questions and open discussion are to
be encouraged but kept to the very end of the session. It is

important that the key speaker gets to present the central message before others attempt to redirect the subject. Interruptions must be handled firmly and the proper response should include a commitment to address the individual question or concern. Those who do have questions can be advised to hear the presentation out, since their questions may very well be addressed by what is yet to be said. A dry run of presentations is mandatory and includes simulation of what might be expected from the audience, up to and including heckling. One should not be overly optimistic as to expect any conversions. The best that one can hope for is that the message is given a fair hearing and heard without confrontation.

There are three specific special interest groups to consider when developing a plan: labor unions, small business, and environmental groups. These groups have been so significant in their impact on project implementation that the next section will be devoted to their special interests and influence. The reaction of these groups to the project may be anticipated. It is important not to be complacent or take them for granted.

Labor Unions

The recession of the early 1980s aggravated an already decreasing influence of labor unions, particularly those in the construction industry. The numbers represented by unions decreased in real terms. The construction effort performed by nonunion contractors has increased dramatically. In addition, several states have joined the right-to-work movement by eliminating the requirement to pay dues to a union to hold a job. Despite these changes, several areas of the country remain strongholds of unionized labor. These are the Northeast and the rust belt of the Middle West. The main differences between union and nonunion labor which provoke labor-management conflict are in the area of hiring and work rules.

Union contracts provide a monopoly to the union in respect to hiring by the employer. The employer may hire whomever he chooses, but the employee will have to join the union. Most construction industry contracts provide that the employer will obtain his labor through what are called union hiring halls.

The most restrictive conditions in union contracts are those covering work rules and work assignment. These provisions define in detail the jobs a particular union is to be assigned and a mechanism by which disputes are to be settled. These are often jurisdictional. Work stoppages due to jurisdictional disputes are now prohibited, but final resolution of assignments can often result in litigation. Compounding this problem are the contracts many large employers have with the national office of a union, wherein job assignments are in disagreement with local practice.

New technology and products not covered in any agreement add
to the confusion.

Several techniques can be used in planning for a project involving employment of unionized labor. Major projects have attempted and often succeeded in obtaining "project agreements" with
labor unions having jurisdiction in the area. These agreements
contain no-strike provisions and wage structures for the duration
of the project. Concessions are usually given to achieve the labor
peace these agreements are intended to provide. These include
slightly higher wages, guarantees of union control over hiring and
discharge of employees, and dispute settlement as regards job
assignment. Another vehicle is the "site agreement," which may
cover detail items such as time of work breaks and time paid for
rain and cleanup. The objective here is to obtain general agreement on site-specific factors which will be applied to all of the
unions represented on the particular site.

One method found effective, particularly on jobs employing
new technology and equipment, is the prejob conference. A presentation is prepared for representatives of the unions with which
contracts are in force. The presentation contains a description
of the project and the proposed assignment of the various parts
to the respective trades. The meeting is scheduled well in advance on neutral territory, such as a hotel conference room. The
presentation is made and the discussion opened on any problems
with any of the proposed assignments. The presenters are prepared to describe in detail the various elements of the job so the
individual union representatives can judge the basis of the assignment to a particular trade. There is rarely unanimous agreement
on the assignments, but it does narrow the disputes for further
clarification and, hopefully, agreement before the work starts.
An atmosphere of cooperation is further enhanced by a social period after the presentation. This type of approach helps break
the ice, allows the parties to meet in a relaxed atmosphere, and
addresses potential problems without disrupting the work itself.

Small Business

The key concern of the small businessman is how the megaproject
will impact his business. This question is universal but for the
owner of a local business, it is a more immediate and tangible concern. The small businessman, whether organized or operating independently, is an opinion leader in his neighborhood and among
his customers. Planning to obtain support for the project must
carefully consider the cultivation of this important segment of
the community.

The association with local business can be beneficial. The project benefits by having a ready source of local supply and reduced

warehousing needs. The local business benefits by the increase
in volume. There are, of course, risks in this arrangement which
must be considered. There is the possibility of higher price resulting from higher demand and limited supply. The introduction
of a middleman can result in delays and stockouts. In the event
of shortages, older and more stable clients will be favored. The
increase in demand may require an investment which may not be
recouped. The anticipated business may not even develop.

There are no guarantees of increased business, and such guarantees are to be avoided. The corporate sponsor of the megaproject gains the most from favorable public reception of the project.
He can neither commit its contractors to support local business
nor guarantee the project will increase their volume. He can communicate the anticipated needs of the project and, when available,
the names of the contractors selected to perform the work. Contractors can be encouraged to support local business. It should
be clearly understood that the company is not acting on behalf of
either the local business community or its contractors. Communications with local business must not result in unjustified optimism
as to the benefits which may be forthcoming. This is particularly
true when the company is in no way able to fulfill implied or perceived promises. The situation is different for operating needs.
The operating company can make commitments for its supplies and
should weigh the benefits of making those commitments with local
business.

6.1.3   Environmental Impact

Few subjects have captured the attention of vast numbers of the
world's population in the period following World War II as that of
the environment. Environmental groups, with a multitude of special interests, are well established and their growth continues unabated. They are now an organized, effectively coordinated and
well-funded powerful force. They must be acknowledged when
their interests are affected. There is no doubt that many of these
groups have accomplished much good. In many instances, however,
an inordinately powerful minority has succeeded in thwarting the
desires of a less coordinated, ineffectual majority. An an example,
the antinuclear interests have almost single-handedly brought a
halt to nuclear power plant construction in the United States.

The risk is underestimating the power and impact of environmental groups, however small. They invoke highly emotional issues on which it is difficult to be associated with the opposing
view. A local issue can rapidly become a national cause. The common characteristics of environmentalists are the intensity in which
they pursue their cause and their missionary zeal.

It is difficult to find an issue, like that of the environment, in which the views are as radical and as polarized. Taking sides in this battle is certain to get your project slowed down or stopped and the costs to escalate. The safest and most effective approach is to acknowledge the power of the environmentalists and work toward neutralization as opposed to conversion.

No project can be environmentally sterile. Someone will find an area in which any project affects the environment. Most states and the federal government require an environmental impact statement prior to siting approval. It is important to address this issue very early. This may even be before the actual site is selected. In this way, there is little of substance for anyone to attack. The company should be prepared to identify impacts and to undertake positive and effective countermeasures. If prepared to go further under pressure, get there first and steal the initiative. Everyone cannot be satisfied and this should not be your objective. Polarization on these issues is extrememly strong and to satisfy the extremists can only be accomplished by dropping the project. If you can achieve the point at which the majority are not fighting your proposal, you have probably achieved the maximum attainable. Taking a point from the earlier discussion, support from the local community will make the job far easier. If those most impacted are favoring the project, outside forces will have difficulty finding supporters to rally.

The focal point for support or determination of who and what the problem areas will be is the environmental impact statement. The states and the federal government require the preperation of formal documents outlining the impact the project will have on the environment and how the project will treat issues such as normal waste, hazardous waste, atmospheric pollutants, plant evacuation in emergencies, and even regional evacuation in the event of a nuclear incident. Committing the company to a position in a public writing before testing public reaction is a sure way to polarize the issues and find yourself behind the proverbial eight ball. Neither can you allow special interest groups with no financial risk to dictate the solutions to environmental problems. What can be done is to determine the positions of these groups, assess the validity and impact of their arguments, and attempt to devise several solutions for each problem which allow for compromise.

The Alyeska pipeline project presents a classic case of lack of acknowledgment of the impact of environmental groups. The pressure of time, as well as the outcry and public support for special interest groups, forced the project's owners to acquiesce on many environmental items, at great expense to the project. Fortunately, the rising cost of crude oil presented a much more attractive return than originally used in the economic justification for the project. This may not be the case for your project. Therefore, the cost of compliance with environmental protection requirements and the reconciliation of problems with special interest groups must be factored into the estimate of cost and time for every project.

6.2   ORGANIZING AND STAFFING

As projects get larger and more complex, we are beginning to look more at organizational factors. In the large-project environment, people become more important. Larger projects are more labor intensive. Because there are more people, we must delegate more, and we must manage people better. . . . With larger, more complex projects, the project manager and his staff, who have always been key figures, acquire much more responsibility: they must be innovative; they must be entrepreneurial; and they must be leaders. [5]

Albert J. Kelley

Organizing for the megaproject must give recognition to its duration and the added dimension of the nontraditional content. Many years may pass between the early planning stages and project completion. The transitional nature of the project over this length of time creates a changing emphasis on various elements of an organization's structure. The increasing adminstrative load imposed by legislation, political and special interest group interface, legal action, and public awareness has created a project within a project.

It used to be that halfway through a project the participants would begin to wonder about what was in store for them on the next project. Now, the concern is whether the project or a career would end first. Nowhere is this more true than with a megaproject. From the time of preliminary planning, it may be a decade or more before the project is completed. In addition to the normal planning and implementation cycle, a third stage has now been introduced, which is a transitional phase. For lack of a better word, it might be called the development stage. It overlaps the other two.

Stretching out the planning stage, the addition of a development period and the delay in commencing implementation require a new approach to project organization structure. This structure must be compatible with the objectives during each of the three periods. It must also be responsive to the probabilities that the project will be recast into a different form before implementation, may be indefinitely postponed, or canceled. The latter takes on a greater probability when one considers the length of these projects, which stretch over such long periods as to be influenced by changes in the business climate. Consequently, the organization must be strong enough to bring to the planning and initial development the strength and direction needed. At the same time it must have a focus to maintain continuity and yet be objective.

Some of these objectives are conflicting and a rational compromise is necessary. At the expense of some loss of objectivity, focus and continuity are the most critical. This can be achieved by assigning the responsibility for planning, development, and implementation to the benefiting organization. They may choose to call on the resources of other functions within the company. It may be that this is mandated by the way the company is organized. The overall responsibility must, however, rest in one place.

Organizational strength augments the transitional nature of project emphasis. The initial planning effort requires the input of a broad cross-section of the company's most perceptive minds. This is in keeping with the strong foundation necessary to make sound decisions on whether or not to commit huge sums for implementation. As the probability increases that the project will go ahead, emphasis will shift to a development mode. This will call into play some very preliminary implementation strategy, but more important are the consequences of going public. At this point planning shifts to how, when, and to whom the reality of the project should be revealed. The public relations staff and media people are consulted at this point. If they have not already been involved, the legal department becomes an important participant. Nearly every megaproject has suffered the consequences of court actions to stop or delay it. In any case, public hearings are likely to be required and these must be planned.

Figure 6.1 illustrates the degree of input of the major functional organizations during the transitional stages of a megaproject. It is meant to cover most types of these projects; however, each must be considered independently. For example, those megaprojects which are international add the dimension of governmental affairs.

The duration of a megaproject has an influence on the choice of its manager. Although the rule still is selecting the best man available for the task, is there such a man? In this case, probably not, for the following reasons:

1. It is unlikely that a single individual will have the level of experience and skill required in each of the three phases of a megaproject.
2. An individual with the requisite skills and experience in any of the three phases is unlikely to accept a decade-long assignment in the same position. If he does, it is likely he will not remain in it.

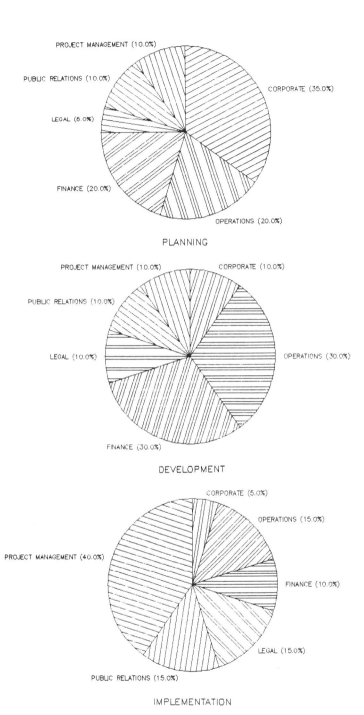

Figure 6.1   Functional input during various stages of a megaproject.

The solution to this dilemma is that a megaproject be managed during various periods by different individuals. There are significant
benefits to be gained from this approach. It increases the possibility of finding suitable candidates. At a point where the project
may still be questionable, a senior executive need not be committed
to a project that may never be executed. Normal personnel development can be maintained. In addition, opportunities for advancement of subordinates are enhanced. The major drawback of this
scheme is an apparent lack of continuity. However, the chance of
continuity of leadership over such a long period is unlikely to start
with. Also, if the company has provided for the natural development of its key personnel, the former project manager is likely to
still be available for consultation. Planning a normal transition is
preferable to responding to disruption.

What is true at the top is also true in the subordinate positions.
The same consideration is required of personal development and
training. Like any growing organization, the best qualified should
fill the positions needed. As the project emphasis shifts, jobs open
which provide an excellent opportunity to reward, test, and develop personnel. The staffing of the project must not exceed the organization's capability of absorbing it once completed. This builds
confidence on the part of the project staff and avoids the loss of
critical personnel as the project winds to a close.

6.3   CONTROLLING

The size and visibility of a megaproject demand controls in addition to those employed in the area of cost and schedule. The image
such a project reflects should be in concert with that of the corporation. It will be perceived as a manifestation of corporate values
and culture. Top management must be aware of critical elements of
the project, yet not involved in its day-to-day management. Controls must be in place which will measure the reception or reaction
to events, compare them to the desired result, and prompt reaction.

The megaproject becomes a public project regardless of who
executes it. Its impact reaches all levels of the economy and hence
has a commensurate visibility and public interest. Running a project in a fishbowl means that the project is seen as a direct reflection of its sponsor. The image reflected by the project team and
its responses to the public must be consistent with the objectives
and culture of the company. They must be ready to address issues of concern in a positive way: that is, to take the initiative
with a program which identifies the project with responsive corporate goals and, if things go wrong, a prompt, effective, and
open response.

Stress induced by change and the frustration of being unable to cope with the perceived threats to safety, the environment, and the economy create a climate ripe with challenge and opportunity. If the corporation ignores the opportunity to calm its public and relieve its frustration, it will find itself reacting to those ready to take up the gauntlet. Their objectives may not necessarily be complementary. The ramifications of either of these consequences should convince top management that it must become involved, stay involved, and in control.

Management involvement begins with reaffirmation of corporate objectives in the development of the strategic plan. Reinforcement comes at the point of review of the plan to enlist public support. Active participation in presenting the corporate and project objectives to its public reconfirms its interest. Reaction, as feedback from surveys or informal contacts, provides an indication of how effectively the plan is being implemented. As suggested earlier, one cannot expect to achieve complete success the first time out or to rest on one's laurels. The message needs to be repeated often and in a variety of ways to provide reassurance and reconfirm the company's dedication to its objectives.

Nowhere are controls more important than in avoiding or handling a crisis. Steven Fink defines a crisis as any prodromal situation that runs the risk of:

1. Escalating in intensity.
2. Falling under close media or government scrutiny.
3. Interfering with the normal operations of business.
4. Jeopardizing the positive public image presently enjoyed by a company or its officers.
5. Damaging a company's bottom line in any way. [6]

Control begins with preparation. The existence of a crisis plan will confirm that management recognizes that a crisis is a realistic possibility. Preparation is necessary to avoid the negative consequences of this eventuality.

Planning for crisis is similar to designing a process system to react to the possibilities of variation of inputs. The objective is to detail the possible situations that pose high aforementioned risks, the alternatives, and who has the action. A corollary is what to do and who should do it when the situation is in doubt. It is like having a fire department: one hopes that its services will never be needed. It provides a level of security: if something does go wrong, a planned response is ready.

6.4   INFRASTRUCTURE

Another factor that distinguishes megaproject from most others is the unusual absence of any infrastructure. Many planners,

particularly those with purely operational background, visualize
the project in its completed form. This myopic view fails to consider the needs required to put the facility in place. As has so
often been the case, the megaproject is in that location best described as "the middle of nowhere." In many instances, an entire
infrastructure is necessary to support the implementation of the
project. This added dimension has a significant impact on the total
cost and timing of the megaproject.

Alyeska and the North Sea oil development, new cities such as
Brazilia in Brazil, Yanbu and Jubail in Saudi Arabia and the massive hydroelectric schemes in Labrador, Venezuela, Uruguay, and
Egypt are but the most familiar of those megaprojects executed in
remote areas. Implementing these projects were projects in themselves and challenged their planners' skill and ingenuity in overcoming the absence of even the most basic needs, needs that many
would have taken for granted and overlooked. No matter what the
project, or where it may be located, a check of these needs may
expose opportunities and illuminate potential pitfalls. These needs
fall into several categories:

1. Utilities
2. Transportation
3. Communications
4. Housing

Each project has its own peculiarities with respect to these needs,
but there are common elements which permit some general comments.

6.4.1   Utilities

Utilities encompass water supply, sewage and waste disposal, power,
heating, and air conditioning. If there is one general rule to consider when planning support utilities for a project, it is to modu-
larize. In the early planning stages it is difficult, if not impossible, to predict the total requirements with any degree of accuracy. Implementation will begin before the full needs may be known.
Small service modules are more readily available and can be installed
more rapidly. If a greater need does not develop, the investment
can be reduced. For example, if peak power demand during construction is expected to be 5 megawatts, this can be provided by
five 1-megawatt installations. They can be purchased incrementally
as implementation expands. Purchase options would enable adding
or canceling units as needs become better known.

The modularization concept carries with it a measurable premium in cost. On the credit side, it will permit earlier implementation of the initial needs. If the full need has been overestimated,
procurement and installation of the last modules may be eliminated.
The installation of several smaller modules provides for a reduced

level of services as opposed to total outage in the event of equipment failure.

Early installation of permanent facilities is always an economical alternative, if these will satisfy the needs of the construction
period. This is often true with power requirements, where the
permanent power needs far exceed those which may be required
during construction. The economics of early spending needs to
be worked out against the cost of installing and operating a temporary system. The permanent system can be an economically viable alternative, if it can be installed in time.

The design of any temporary facility should take into consideration its impact on the environment. Temporary sewage treatment
should meet or exceed the effluent standards which will be prescribed for the permanent installation. Solid waste should be disposed of in a well-designed and well-operated sanitary landfill. If
there are any hazardous wastes, these must be disposed of in the
manner prescribed by law. In the absence of adequate laws in
many foreign countries. U.S. law can be used as a standard of
compliance. Remember, the construction phase of the project is
the first opportunity its permanent neighbors will have to see if
the company's deeds live up to its words. These first impressions
will have a lasting impact and they should be good ones.

Water systems require particular attention. There are probably
few places where a major project is constructed that water of some
quality is not available. In those rare instances, water would have
to be hauled and carefully rationed. Given the quantities required
and risks involved, water must be treated at the point of use. The
choice is whether to treat the entire supply to be potable, or to
provide a dual supply. This may be a choice of pure economics,
but several things must be considered in making a decision. In
many foreign locations, particularly in Third World countries, custom and lack of knowledge prevent a distinction between potable
and nonpotable water. Whatever comes out of the pipe is assumed
drinkable. Establishing a dual supply, including the special features necessary to prevent accidental drinking, may be far more
expensive and a greater risk than treating the entire supply to be
potable.

Heating and air conditioning are mentioned as a reminder that
in many locations these services may both be necessary. If not
necessary, they may at least be desirable. If they are provided,
they will impose a load on the temporary power system and create
the needs to handle and distribute fuel.

6.4.2   Trasportation

Remote sites, innovative labor-saving opportunities, and the size
of plant components provide unique challenges in megaproject
logistics. This includes both the means and modes of transportation.

Modularization, as a means of reducing on-site labor, has been a common element of many megaprojects. Transportation of outsized and extremely heavy modules requires careful planning for their fabrication and conveyance.

The high cost of on-site labor in remote areas has led to a high degree of refinement in the modular concept. Modularization is nothing more than the preassembly of component parts of a plant location. What started out as the preassembly of rotating equipment and its driver was extended to the preassembly of entire plants. Floating power and water treatment plants have been built and operated in many remote areas.

Specially designed carriers are available to handle outsized and heavy cargo. In many instances these were built for a special purpose at the time and their availability then led to designs built around their capability. These are ocean-going vessels which are capable of transporting and discharging multiwheeled dollies loaded with a plant module. There are self propelled multiwheel transporters which are also self-leveling to prevent stress on the preassembled module. Several unique vessels have been built which can submerge below a floating module or barge and then rise to lift the load. At the destination the process is then reversed. Derrick barges have been built for construction of North Sea oil platforms which have capacities in excess of 2000 tons. Innovative techniques and equipment seem limited only by the financial justification required to implement them.

When considering modularization, it is important to remember that although one can be transported halfway around the world, it's the last few feet that count. Local restrictions, narrow roads, bridges, underpasses, and site conditions all must be considered and accommodated. This may mean building additional roads, constructing bypasses, shoring bridges, or cutting and splicing power and telephone lines. The additional costs of these efforts must be factored into the decision equation.

A key element in the transporting of unusually sized equipment is timing. Although prefabrication of modules or the factory completion of long or heavy process vessels may be cost effective, this advantage may be lost if they cannot be transported as planned. The number of special carriers is limited and bookings must be made well in advance and maintained if the savings are to be realized. Adequate planning and careful control of the fabrication process are necessary if all of the events are to come together at the prescribed time.

6.4.3   Communications

There are two facets to project communications: those within the project and those of the project and its outside interfaces. The remoteness of most megaproject sites adds to the problems associated with both. Communications on the site itself is a boot-strap operation. There is often nothing in place on which to build and a system must be created from the ground up. External communications problems are compounded by the local problems. These may be magnified by governmental restrictions or limitations in available systems. There is a balance between radio transmitters and telephone communication which must be examined for each specific site. There is also a general tendency to overemphasize the benefits of instant accessibility. Expensive radio equipment is difficult to justify by cost benefit analysis. With the exception of security, medical, or fire brigade personnel, portable radios and pagers take on the image of status symbols. If employed, their usage must be strictly controlled and monitored for confirmation of the benefits that justified them.

Strategically placed telephones provide an excellent, inexpensive, timely alternative to the radio. In many areas of the world, particularly the Middle East, the radio is looked upon as a useful tool for those opposed to the present government and is prohibited, or severely restricted. Obtaining approval to import radios and administering the required controls significantly add to their cost. Actual usage studies on remote area pipeline projects have demonstrated that, even under conditions of remote and widely dispersed construction locations, the actual need and usefulness of radios are very low in comparison to the desire to have one.

External communications are limited by local capability and are beyond the control of the project. Local service capability never anticipated the extra volume, scope, or sophisticated requirements of the project. It is safe to assume that response to project needs will be less than desired. If the project is to have the required communication with the outside world, it must supplement local capability, establish independent means, or both. This is often accomplished in an environment of skepticism, bureaucratic delay, and, on occasion, obstruction.

Starting early and being armed with well-developed justification will help in overcoming external communications problems. A plan must be developed before approaching local authorities. It is based on an evaluation and assessment of the local situation and the plan is to obtain effective communications. The effort boils down to selecting the appropriate strategy. The central problem is a significant dependence on an outside agency over which there can be little control. If the local authority has a good record of response in a timely fashion, it is probable they will accommodate the request. Often, the results are merely promises. This is due to lack of funds, staff, and management to address any new short-term demand. An offer to take over the problem becomes an affront. If carried too far, it may result in official obstacles

preventing independent action. An effective approach is an informal, low-key offer to assist, with a promise to turn over equipment to the local authority for their use. This could involve satel-
lite dishes, telephone exchanges, or microwave equipment. As expensive as this may sound, the ultimate cost of needed additional
equipment will be borne by the project in any case.

6.4.4   Housing

Starting a megaproject is probably going to put the project's sponsor in the housing business. There is either insufficient housing
available, or it is located too far away. Problems at the worksite
will pale in comparison to those experienced in the area of housing
and its associated catering recreation. No matter what class of
facility and service is provided, it will not satisfy everyone. Nevertheless, planning must include providing clean accommodations,
wholesome food, and reliable services and concentrate on who should
provide them and how.

Running a temporary hotel for large numbers of isolated ba-
chelors is certainly not a business in which most companies want
to engage. Unfortunately, most megaprojects make this alternative inevitable. There will definitely be some hard choices. One
of the first is whether to take on this task as an owner obligation
or leave it to individual contractors.

Keeping in mind the previously stated objectives, assuming responsibility directly ensures that action can be taken toward achieving these objectives. If the responsibility is to be delegated to
individual contractors, the objectives can be stated in tender documents. How well this contractual obligation is carried out depends on the contractor. No two contractors will perform this task
the same. If contracts are on a fixed-price basis, the pressure
exists to provide as little as possible. An owner has to assess if
these two consequences are in his best interests and how high the
risks are.

Several factors work in the owner's favor when the choice is
to turn the responsibility over to contractors. If there are a large
number of contractors working on the same site, the tendency will
be to establish a standard of housing and services close to the
higher end. The risk to contractors in not making this shift is
the loss of staff to the competition. The high cost of recruiting
and turnover is usually enough to force a reluctant contractor in
this direction. If the workforce is all from the same general background or nationality, expectation will generally be the same, and
arriving at a standard between contractors is much easier. If the
labor force is unionized, as was the case on the Alyeska project,
the union adds a powerful force to individual complaints.

Many projects are carried out in countries where labor has to
be imported. Standards will be very different from those in the
industrialized West. It is the expectation of the imported labor
force which is the critical concern. Morale and the attitude of labor
toward the project sponsor will be poor if what they find in working on the big megaproject is exactly what they had on the last job.
The varying standards between contractors from different countries
is also likely to cause friction. The pressure to provide as little
as possible under the terms of fixed-price contracts and the relative ease with which contractors can accomplish this almost ensure
it will happen. Local contractors wield tremendous power over their
labor. It is rarely organized and depends entirely on the contract-
or for a livelihood. The alternative to accepting whatever is offered
is usually unemployment. Contract provisions notwithstanding, it
is difficult to enforce compliance and canceling contracts is an unacceptable alternative.

As a general rule, if the project is in a developed country and
the labor is organized and from the same general background, housing and services can be left to the contractors. In all other cases
the owner should consider various levels of involvement. As megaprojects have developed, so have contractors who specialize in providing this type of service. Contracts for this service should be
specific and detailed.

In 1975 the Saudi Arab government charged the Arabian American Oil Company (ARAMCO) to execute a multibillion-dollar program,
which came to be known as the Government Gas Program. It involved the collection, separation, liquefication, and shipment of oil-
associated gas which had formerly been burned. The number of
people directly engaged in construction exceeded 35,000 at its peak
and involved contractors from all over the world. Before a single
construction contractor was selected, a program was initiated to
house, feed, and provide medical and recreation services for this
workforce at six separate locations. Also included were centralized
office complexes complete with services for communication and blue-
print production. There was even a separate complex at two locations for up to 400 families. Justification for this investment was
based on the following factors:

1. A single standard for all contractors could be provided
    and maintained.
2. Lead time for developing and selecting contractors could
    be considerably shortened. (It was correctly assumed
    that few, if any, contractors had the facilities to house
    the workmen they would require to perform the work.
    They would not purchase such facilities in advance of
    having firm contracts and hence would require additional
    lead time to mobilize.)

3. The final cost would be lower. This was based on the
    cost of contractors to maintain the required standard and
    writing off the total cost against each contract. It was
    assumed that capital and operating costs to ARAMCO would
    be lower due to volume purchasing. The savings were es-
    timated at 25%.

Despite some early setbacks, the objectives were achieved, if
not fully provable. Several years after this program was implemented the Saudi Arab government chose to emulate this program
when they began development of the two new cities of Jubail and
Yanbu.

Some years later new contracts provided that contractors would
be responsible for housing and feeding their employees. At one
location where about a dozen contractors shared a common site,
health and sanitation violations were common, as were outbreaks
of stomach and skin ailments. In one case, men were housed in
shipping containers, with no windows. In another offshore location, 250 men were crowded on a converted oil tanker with three
toilets. Incidents of health and sanitation violations and breaches
of the peace in just one of these locations far exceed those of the
entire earlier program. The initial contract cost was less, but the
ultimate cost significantly higher.

6.5 SUMMARY

The megaproject has a significant dimension in addition to those
normally identified in the routine project. This is the importance,
significance, and risk associated with external interfaces which are
added or amplified by the magnitude and visibility of the megaproject. It is essential that the planning effort acknowledge this difference and include positive action to manage this aspect of the
project. The information program must be developed and presented
in such a way as to take the initiative in obtaining the support of
the community in which it will be located, its leaders, and interested parties. Those who are likely to oppose the project must be
identified and the forms of their opposition determined. Positive
measures must counter misinformation and rally supporters to neutralize opponents. Special interest groups can be enlisted by identifying project benefits in which they will participate. Aspects of
the project that may have a perceived negative impact, particularly
those involving the environment, must be identified. Positive measures to be taken to minimize them must be identified and implemented.

The special nature and duration of the megaproject require a
novel approach to organizing and staffing. Each phase of the

project has unique requirements and involves different skills and experience. It is advisable to consider having a different leadership for each phase. The duration of the project places special emphasis on methods to maintain normal personnel development. The different phases present opportunities to reward performance and provide a variety of experience.

The magnitude and visibility of a megaproject represent a condition of high risk. This mandates involvement by its sponsor's top management. Controls must be in place to maintain rapid dissemination of information on status and problem identification. Situations that may bring unfavorable publicity and adverse public reaction must be identified immediately and countered effectively by prompt and positive action.

Megaprojects are most often executed in areas remote from any infrastructure or support. This adds a significant factor of time, cost, and challenge to the problems of implementation. Uncommon areas of effort can result in items being overlooked or underestimated. Planning and thorough review again are the cornerstones on which effective execution is built. The scope of what is required may be broad and problems compounded by any lack of capability or capacity or indifference by the local authorities to support the project may further add to the problem.

Megaprojects stand as a monument to man's ingenuity and determination in the face of interminable odds. Of the many that have been completed, few are of the dimensions in cost and time as when they started and yet their proponents persevered. Their misfortunes and setbacks have not discouraged the attempt of others. At this writing, work has actively started on a 200-year-old dream, a tunnel under the English Channel. Although probably not the most costly, this is certainly one of the most exciting projects of the twentieth century. More are undoubtedly being considered and on the drawing boards. Their proponents would do well to heed the lessons of the past.

REFERENCES

1. Albert J. Kelley, *New Dimensions of Project Management*, D. C. Health & Co., Lexington, MA, p. 141.
2. John C. Dvorak, Inside track, *PC Magazine*, p. 79, Oct. 14, 1986.
3. Peter F. Drucker, *Management Tasks and Responsibilities*, Harper & Row, New York, 1974, p. 472.
4. F. L. Harrison, *Advanced Project Management*, John Wiley & Sons, New York, p. 5.

5. Albert J. Kelley, *New Dimensions in Project Management*,
   D. C. Health & Co., Lexington, MA, p. 11.
6. Steven Fink, *Crisis Management*, American Management Association, New York, 1986, p. 15–16.

7

Managing the Troubled Project

The troubled project occurs more frequently than imagined. The reason that most are unknown is that their problems are uncovered promptly and corrected. The troubled project is burdened with sunk costs, in the form of a poor plan, errors in judgment, and errors in action. The sooner one recognizes these sunk costs, the sooner one stops searching for the culprits and gets on with the positive actions necessary to evaluate the situation and take corrective action.

The project, as a short-duration phenomenon, is no different than other kinds of emergencies which require one to put out the fire or stop the bleeding. It demands a prompt assessment of the situation, immediate application of measures to stabilize the status quo, and implementation of actions necessary for correction.

The troubled project gets that way because the people closest to it fail to recognize system or personnel faults that create them or are powerless to correct them. This means that an arm's length oversight system must be in place to monitor project status. It must be consistently employed and proven capable of detecting system and personnel errors.

Often an analysis of project problems leads to the conclusion that the project itself was ill conceived. Hard choices as to whether or not to continue the project need to be made. Terminating a project before its completion is always a choice and often the best one, but it takes courage. It should not be one left to the project's management, but the inability to detect problems and ~~pro~~ delayed ~~erastinating in~~ their correction may result in just that.

As in any other team effort, when the results expected are not being achieved, the consequences of failure are laid at the feet of the manager. Replacing the manager is a solution most often employed. If done because the troubled project got that way because the manager failed although he had the people, tools, and systems to succeed, then it is the right choice.

7.0 INTRODUCTION

A project is very much like a human being. It has life, requires care and maintenance, occasionally becomes ill, and eventually is terminated, sometimes prematurely. Even with proper care and maintenance, it is prudent to have an occasional checkup to detect the early warning signs of potential problems. Illness comes to all regardless of attempts at avoidance, although usually more severe in those less prepared. Prompt and decisive action often brings a speedy reversal. Procrastination and failure to impose the required regimen result in prolongation. The medical parallels from a diagnostic and prognostic point of view are so striking that they provide a natural format in which to tackle the subject. For example, there are congenital problems and certainly euthanasia.

7.1 PREVENTION

It is better to prevent trouble than to recover from its consequences. Prevention begins with recognition that no system is perfect and periodic, objective assessment of project condition is essential to overcome this inherent weakness. A well-designed system is still a basic necessity, the key being that an effective system will reveal those subtle changes that provide an early warning of deviations which require attention. Correction at this point can be modest, inexpensive, and nondisruptive.

Prevention methodology should also anticipate that the control system may not be accomplishing the objectives for which it was designed. This is particularly true when projects are novel efforts or executed irregularly. The most effective means to overcome these inherent weaknesses is early and frequent peer and management review. As confidence builds or corrective measures take effective hold, the number and frequency can be reduced. Modification and regular updating of the control system and its data base are essential, even in well-designed systems.

7.2   DIAGNOSIS AND CURE

> Bad news is like bad food, the longer it sits the worse
> it tastes. . . . Very bad news takes time to develop,
> to ferment and to grow. More often than not, the causes
> of information shock have been at work for months or
> even years, we just didn't know about them. [1, p. 159]
>
>   Robert D. Gilbreath

Signs of impending trouble or even its presence are of little
use unless one can diagnose the problem correctly. Often the apparent cause is not the real cause. Although the problem may manifest itself in a cost overrun or schedule slip, the cause may be
complex and involve both. It can contain elements of project mis-
administration or poor technical performance. Examination of even
the healthy project should be thorough and complete. The basis
for project success or failure is the foundation on which it is built.
It is this foundation which requires close and frequent scrutiny.
Like any good medical examination, diagnosis begins with an understanding of the anatomy. The environment in which it functions and the hazards that environment presents coupled with inherent weakness will determine what may go wrong. It is on this
basis that prevention is undertaken and, failing that, steps toward
a possible cure. The examples are meant to illustrate that the first
effort in turning around a failing project is identifying the real
problem. This means knowing what should have been done and
then ascertaining what has been done. This takes experience,
knowledge, and a lot of patient digging. The deeper in trouble
the project, the harder the digging. There is no easy method or
simple approach which says: if this is wrong, this is why and this
is the cure. If it were that easy, there wouldn't be any failing
projects.

7.2.1   Cost

Examination of the cost foundation is a relatively straightforward
exercise and, if done with care, yields an early warning of impending problems or builds confidence. It is entirely possible that
the project was underpriced. The overall cost basis is fixed. Depending on the structure, it provides an accurate measure of variation. Although most cost elements are not directly controllable,
recourse as to their impact is available due to the nature of the
cost continuum.

The working budget is the basis of cost control. It provides
the ultimate dimension. The estimating elements are the blocks

from which the whole is constructed. The greater the number of individual blocks, the easier it is to develop a clearer picture of the whole and the quicker it comes into focus as development progresses. Successful projects have been executed within a prescribed budget when many of these building blocks are uncertain. Conversely, many projects have gone sour when the budget was supposedly built on highly detailed estimates.

Most organizations utilize various estimating levels. It is not important how many levels are used, but to know how each level is determined. The most common are shown in Table 7.1. The higher the degree of inaccuracy, the more important the need to define the basis of the estimate. Despite the level of inaccuracy assigned to the estimate, judgment is often rendered on the decision maker's perception of the project as defined. Therefore, the estimator should always proceed on the basis that if funded, the project could be completed within the estimate and the accuracy indicated.

Many organizations temper the estimating accuracy by including a minus factor. This is misleading. The more important consideration is whether the project is profitable or viable in the event its cost may be higher by the indicated percentage. As a practical matter, the reason the estimate is inaccurate is because insufficient definition is available to cost all of the elements. It is highly unlikely that better definition will exclude elements and reduce the probable end cost. There is a probability that individual elements in the estimate can be obtained at less cost. This is true at all levels of accuracy but until the project becomes a funded effort, the important factor is the possibility that the cost could be substantially higher. Nothing should detract from this element of the image the estimator is trying to convey to management.

The estimate of cost of individual items is comprised of several elements. The first is the base cost, the price last paid or quoted for the item. To this is added an escalation factor to arrive at a current price. To this sum is added an element to provide for changes resulting from requirements as design develops. This may be large, small, or zero, depending on the item and the probability that it will be impacted as design evolves. To this sum is added a contingency to account for errors in estimating any of the aforementioned factors.

As the project progresses, the opportunity arises to assess the criteria used in developing the estimate. This involves examining in detail the actual base cost, escalation factors, change, and contingency allowances. The first would be the engineering design contract cost and the long-lead major equipment elements. These are important because they represent significant cost

Table 7.1 Estimate Type, Accuracy, Basis and Use with Sequences of Activities Occuring over Time

Estimate type	Accuracy range	Basis	Use
Order of magnitude	40%	Cost per unit area volume or production rate from previous experience or commercial estimating manuals	Evaluation of projects to include in long-range budget planning
Budget	20%	Process flow diagrams; plant office, or building layouts	Determination of projects suitable for additional development funds
Appropriation	10%	Fully developed project proposal and detailed scope	Selection of projects for implementation
Control	<10%	Firms bids for material, equipment or construction	Project cost control
40% Order of magnitude	20% Budget	10% Appropriation	0% Completion
Project definition	Planning	Preliminary design	Execution

elements and provide the first opportunity to critically assess the estimate. From the standpoint of timing, this also represents the first point of no return. It is at this time that the project can still be abandoned at the lowest cost.

The cost of the design engineering effort is developed from the organization chart of the design team and the services to be provided. The base cost is the number of personnel, their hourly cost, and the duration of their assignment. To this is added services such as reproduction, computer time, communications, and other related costs not included in the hourly rates. An overhead percentage is applied and a fixed or variable fee, depending on the type of contract. On an engineering contract a percentage can be used to account for potential changes. A contingency is then applied to provide for uncertainty.

If a detailed estimate of the engineering cost is missing in the budget estimate, these elements become available when proposals are solicited for the engineering contract. The request for these proposals should contain a requirement to provide this sort of breakdown. The proposal then enables a check on the gross estimate and provides the detail required for subsequent control.

Proposals for engineering contracts provide confirmation of estimate accuracy in several important areas: total effort as dimensioned by the gross man-hours required, wage and salary rates of professional personnel, cost of services, overhead, and profit rates. Projections of estimated final cost can be made using this information, but should be assessed with caution. For example, the variation between the anticipated cost of engineering and the revised cost based on the actual proposals may be applied to the total cost. On a selective basis, variations in man-hours and overhead estimates may be applied to similar elements in the construction portion of the project. The assumption is trends in assessment and actual cost of the two move in concert. Table 7.2 gives examples of the results of this exercise and assumes that half the contingency is intended to cover estimating errors and half to provide for changes.

A careful examination will show that contingencies have been adjusted specifically for the revised estimated costs. This is intentional since half the contingency was available to provide for estimating errors and was insufficient. There is still uncertainty until the project is completed, but the degree of uncertainty is now diminished based on new information. The overrun of the working budget reflects the original intent of the project to maintain half the contingency for changes. This cannot be met without incurring extra cost. If the estimate accuracy is 10%, the expectation would be for the appropriation of a maximum of $5,720,000. In nearly all cases, the appropriation is the same as the working budget.

If management wishes to maintain the working budget as a rigid control, a reduction in the allowable changes will have to be made to reduce the expected overrun. At this point only the first 13% ($5,200,000/$680,000) of the project's funds are to be committed. Little has been spent, except for internal expenditures on staff. The process illustrates the decision points and the factors to be considered in controlling project funds.

At the same time the engineering contract is being developed, quotations are being received for long-lead engineered items. Assume these include process pumps, compressors and associated drivers, and a process column. While engineers are checking the technical quality of the offerings, information can be gleaned from the prices quoted. Table 7.3 presents a comparison of revised estimates of cost based on the quotations with that in the original estimate and budget.

An examination of the table will illustrate several areas where effective controls can minimize potential future problems. In developing the budget it is assumed that half the contingency provides for estimating errors and the other half for changes. In the case of the vessel, if the change contingency is 5%, when applied to the actual purchase cost, the allowance reduces to $1350. Similar logic applies in the case of the pump sets. This contingency is not quoted by the supplier, but is reserved as an allowance for anticipated changes.

Provision has been made for an alternative to the three 50% capacity unit preferred design. This is an option to selecting two 100% capacity units. This alternative may have been suggested as a means to determine the premium required to maintain the preferred design or as an opportunity to reduce project costs. It is often offered unsolicited by vendors who may not have the exact capacity requirements in their standard supply and must offer a larger unit which must be modified to suit. The alternative is also offered in order to be competitive.

The original three-pump option had been justified by greater operating flexibility and lower operating costs. The cost picture provided presents the project manager with various alternatives:

1. Review the technical specification for the preferred alternative to determine whether the cost can be reduced to meet budget constraints.
2. Rework the operational justification to determine whether the choice is still justified. It may be influenced by the fact the two 100% units are also more costly than originally estimated.
3. Select the lower cost alternative.
4. Proceed with the preferred design recognizing that the total budget for the three items has not been exceeded.

Table 7.2 Projections of Revised Final Project Costs Based on Firm Engineering Proposals

	Budget
Engineering	$ 618,000
Material and equipment	2,364,000
Construction	1,473,000
Other	273,000
Contingency	472,000
Total	$5,200,000

	Company	Contractor	Error%
Base engineering man-hours	$ 19,800	$ 21,000	-6
Average hourly cost	22.10	21.60	+2.3
Subtotal direct labor	$441,400	$453,600	-2.7
Subtotal overhead cost	$176,600	$204,120	-15.5
Contingency 10%	$ 62,000	$ 31,000[a]	
Total estimated cost	$680,800	$688,720	

Error in engineering estimates as applied to construction costs

	Company estimate	Error %	Expected contractor proposal
Base man-hours	43,600	-6.0	46,200
Average hourly cost	26.00	+2.3	25.42
Subtotal direct labor	$1,133,000	-2.7	$1,174,000
Overhead percentage	30	-13	34
Subtotal overhead cost	339,800	-15.5	393,000
Contingency 10%	147,000		73,600
Total estimated cost	$1,620,000		$1,641,000

Revised estimate of final project cost

Engineering	$ 688,720
Material and equipment	2,364,000
Construction	1,567,400
Other	273,000
Contingency	368,000
Total	$5,261,420

[a]Not included in contractor proposal

Table 7.3 Comparison of Estimated Costs of Long-Lead Equipment Based on Quotations and Original Budget

	Budget	Firm quotation	Difference
Pressure Vessel PV-1	$30,000	$29,525	-$ 475
Estimating contingency 5%	1,500	–	- 1500
Estimated changes 5%	1,500	1,475	- 25
Totals	$33,000	$31,000	-$2000
Pump set P-1			
Alternate 1 – 3 – 50% units	$40,000	$43,810	+$3810
Alternate 2 – 2 – 100% units		39,300	
Estimating contingency 5%	2,000		- 2000
Estimated changes 5%			
Alternate 1	2,000	2,190	+ 190
Alternate 2	2,000	1,950	- 50
Totals			
Alternate 1	$44,000	$46,000	+$2000
Alternate 2	–	$41,190	-$2810

If this project had been effectively controlled, the project files will contain: a request to examine the technical specifications and its results, a reevaluation of the pump and compressor selection economics, and a request to buy the preferred units. If the project were in financial difficulty, the alternative units provide a means to reduce it.

Two often-repeated actions can result if adequate controls are not in place and well understood. The first is the substitution of lower-cost alternatives without review by those impacted. This occurs when budgets are tight or overrun and the project manager is under extreme cost constraint. The second, which occurs most frequently, is utilization of the entire contingency to cover an initial commitment, leaving nothing for the future. In the example, applying the savings in cost for the vessel against the higher cost of the pumps is not an unreasonable approach. The problem arises if this option is unavailable. Utilizing the entire contingency for the pumps would enable the working budget total not to be exceeded. This short-term manipulation is all too frequently employed, and when the inevitable changes are required, there are no funds left to be applied to them. As these accumulate, the need for further appropriations arises. When these must be requested they are a fait acompli, in that they must be approved or the project cannot be completed. Leaving management with few, or no, alternatives is a sure method to shorten a career as a project manager.

This example illustrates that even in well-planned and well-organized projects there will be numerous occasions when judgment is required. The degree to which that judgment can be exercised may be totally unfettered, or it may be restrained within bounds commensurate with the experience and skill required for the position. If a project manager operates with a system of adequate and reasonable controls, the proper actions taken in this example will come as a natural course of everyday activity. Manipulation to escape the apparent budget problems or to make points by underrunning the total budget would be considered a lack of judgment.

The example is skewed to illustrate another area requiring judgment — that is, offsetting the lower cost of the vessel, such that the total of the three does not exceed the original budget. The question is who has the authority to exercise judgment as regards the utilization of contingent funds. If management wants to control the risks and hold the project manager responsible for a fixed budget, it will retain control of the contingent funds. In this example, approval would be required to exceed the budget in the purchase of the pumps and compressors, irrespective of the offset available from the vessel. The project manager would

be aware of this restriction and could proceed without approval
only at his own peril. Judgment would have him recommend the
preferred alternate at the higher cost.

At this point in the project, a better picture has come into
view. Original estimates in major areas have been put to the test.
It is the last opportunity to make a significant alteration in the
project or even to abandon it without incurring a substantial cost
penalty. This opportunity should not be allowed to pass without
making this choice.

It is imperative that the project be subject to a high-level review before making engineering or long-lead purchase commitments.
If the project is to continue beyond this point, its basis should be
reconfirmed or readjusted as required.

No budget is complete without a schedule of commitment and
expenditure. The finance organization requires the expenditure
schedule to have funds available to pay the bills. The commitment schedule is an important tool in cost control. Each expenditure is preceded by a commitment. If those commitments are not
made in a timely fashion, the project will be delayed. The tracking of actual versus planned commitments and expenditures provides an early warning of potential delay. A typical schedule is
shown in Figure 7.1.

7.2.2   Schedule

When one hears of disastrous schedule slippage in a project, he imagines that a series of major calamities have
befallen it. Usually, however, the disaster is due to
termites not tornadoes; and the schedule has slipped
imperceptibly but inexorably. [2]

                                             Robert J. Graham

A schedule's most dangerous enemy is optimism. Despite the
lessons of the past, many project managers approach each task
in the assumption that they will do better. Without fail, another
hard lesson is learned. It is difficult to fault this approach because knowing what may have gone wrong before, it could be
more easily avoided the next time. What is overlooked is that no
situation quite repeats itself exactly. There is always some slight
difference, or condition, which defies a direct comparison. When
early activities confirm this overoptimism, they are often ignored.
As with cost, the time required to perform various tasks may have
just been underestimated.

A project falling victim to overoptimism gets heavily massaged.
As time begins to slip away, the project planners begin adjusting
the logic of the sequence of events. This is possible, as no

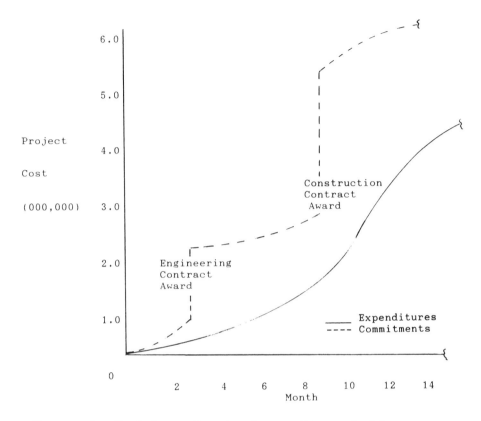

Figure 7.1   Typical commitment and expenditure schedule.

schedule can include an infinite number of activities. It is then
possible to add, delete, or combine activities in a new schedule
and arrive at the desired results. As with the juggler adding
oranges, there is a limit as to how far this can be carried. Before long, everything in the schedule is critical and no opportunities available not requiring costly overtime or double shifting. The
alternative is delay. At this point, major surgery is all that is left.
    There are two methods available which will uncover the pre-
sence of overoptimism. As costs of engineering services and long-
lead equipment items are being examined, their schedule consequences should also be reviewed. Estimates of the man-hour

content of the design task, available resources, and the program of staffing and destaffing the effort will confirm or refute the schedule of the design work. The sequence of design, certified drawings, manufacture, assembly, test, and shipment of long-lead equipment items will do likewise.

A suggested approach to soliciting proposals for both services and special equipment is addition of the option for the supplier to provide his optimum proposal. This was discussed earlier in the section on contractor schedule control. What is desired in the way of schedule performance may not be exactly what the supplier can provide. Money may be saved at the expense of some schedule adjustment which may be acceptable. Unless there are definite incentives for the contractor or supplier, the schedule will be that of the contractor and the possible savings lost.

The other technique to prevent leakage of the float in the schedule is discussed in the chapter on control. The schedule reviews by management should always include the original critical activities that were to have occurred within the interval. Although not absolutely foolproof, examining the current critical activities against the original will indicate the level of manipulation. Examining the slippage, if any, of these original critical activities is still a good indicator of potential problems.

The prescription for schedule recovery is the same at any point in time. The earlier it is administered, the more palatable the medicine. Selective treatment of critical activities is the first step. Overtime may be applied to reduce the duration of these activities to recover lost time. The less labor intensive the activity, the lower the cost. Additional manpower can be applied and double shifting may also be effective. As time and events pass, the options diminish. In the construction phase, it may be impossible to work double shifts and selective overtime because of prohibiting labor contracts. More parallel activities may be going critical, increasing the total number of activities that have to be "crashed" (a term taken from the CPM vocabulary meaning applying effort to critical activities to reduce their duration).

Anticipation of schedule difficulties is a prudent form of prevention. Contracts and purchase orders should always include requirements for the contractor or vendor to recover time they themselves have lost. In addition, critical and near-critical activities should be candidates for bonus and penalty conditions. For troubled projects it is important to recover lost time. To achieve this, additional capacity must be identified, made available, and put in place.

7.2.3   Administration

The non-technical causes [of project failure] are not always so evident, sometimes more insidious and less

easy to apply or avoid correctly on future projects.
This is further complicated by the fact that technical
and non-technical people feel competent to discuss
non-technical issues. [3]

In nearly every situation where something has gone wrong,
the first action of the postmortem is a request for the records.
There are few instances where the records are not as chaotic as
the situation. It is almost inevitable that those who get themselves
or situations into difficulty are also poor record keepers. If there
is a connection between the two, then bad situations could be prevented from developing by routinely and accurately documenting
the activities that must take place.

Good documentation also means standardization of approach.
No matter who acts or where an action is taken, recording and reporting it should be consistent with a prescribed format. The objective is to be sure the only difference between projects is their
scope and schedule, not how they are administered. In this instance a procedures manual is necessary to achieve the desired results. The major categories of documentation and records contain
at least the following:

Contracts
Correspondence
Costs
Drawings
Estimates
Invoices
Personnel records
Purchase orders
Schedules
Specifications
Standards

Project office setup is a responsibility of the department responsible for project management. Setting up the office means
assembling the drawings, documents, manuals, and standards and
getting them to the project team. This department is also responsible for revising the manuals and standards subject to change and
revision. This method assures the standardized administration of
each project office and that those offices are functioning with the
latest data. In many instances they function as clearinghouses of
information and advice and can serve management as auditor of
the administrative aspects of project execution.

It must be recognized that each project manager will have his
own methods to impose his style and personality on a project. The
rules should be maintained in a simple and flexible manner so as
not to stifle initiative or appear an attempt at cloning. On the
other hand, the system should be such that socialization into the

project can be accomplished in a very limited time. Reviewers of
project activities, as reflected in their records, must also be able
to make their assessments rapidly as though they were traversing
familiar terrain.

Reporting is the most significant aspect of project administration. It serves as a means of communication to higher levels of
management and as a record of project progress. In addition to
the glossy photographs and carefully constructed prose, there
must be the bare fundamentals of project status. If it exists nowhere else, the presentation of the essentials of project condition
must be in a standard format. The reviewer of the reports must
be able to assimilate quickly and accurately what is being communicated without having to determine how. As easy as this would
seem to accomplish, it is by far the most flagrantly violated.

Project reports must be structured to permit the data presented to be rolled up easily into an accurate, informative summary.
This is essential since each level of management that gets the information is looking at it from a different perspective. The need
being served must be accommodated and the superfluous data sup-
pressed. How this is done will vary from organization to organization. It should reflect the responsibilities of the various levels
involved in the management of projects and their implementors.
In organizations such as construction companies or engineering
firms, where projects are their core, a status report and two summaries are all that is required. The reasons become more apparent when examining the major functions and responsibilities of
the various levels involved.

Project Manager – This is the operational level responsible for
the day-to-day execution of the project. Reporting obligations
consist of presenting an accurate picture of project status and
condition. This includes an assessment of opportunities or problems and what is being done about each.

First-Level Supervision – A major function of this level is assessment and evaluation of the performance of the project managers in his charge. This includes a continuing obligation to training and development as well as providing guidance and assistance.
The first summarization of the individual projects would include
budget and schedule data and an assessment of the achievement
of project objectives. If there are problems or opportunities, these
would be highlighted and action indicated.

Second-Level Supervision – Where this is in the corporation is
not important. Its responsibilities for the conduct of projects are
a part of a much broader responsibility. At this level the concern is fiducieary as opposed to an operational responsibility with
respect to project execution. That is, certain goals have been

established and funds have been appropriated to achieve them. Are these goals going to be met within the funds appropriated?

The first-level summary is constructed to answer this question and to decide what is going to be done if the goals will not be met.

As an example, take a project that has five basic subdivisions. These could be distinct areas in a plant project or departmental assignments in a developmental or efficiency improvement project. The project manager's basic controls would include an estimate of the detailed costs for each area and a schedule of the various activities. The report would contain a comparison of costs and events in each area and a reassessment of the final costs and completion dates. As suggested in earlier sections, a critical event would have been planned to occur during the report period. Its status would be reported with any explanation required. This would include updated commitment and expenditure curves. Trends, if any, would be indicated, along with an assessment of their meaning and any action planned.

It is assumed that the supervisor reviews the project with its manager at regular intervals. It is the project staff who condense major portions of the report for use by the supervisor in his report. The area reports are consolidated into a single cost and schedule status for the project. The report might indicate, for example, whether the status and assessment of the project indicated a need for additional funds or disclosed surplus funds which could be utilized by management for other purposes. The report might include a proposal as to the disposition of these funds. The supervisor would then append his own assessment of the status and condition of the project and its management. This report is included with reports of the other projects being executed under his supervision. At this level, a supervisor must have a working knowledge of the project's objectives, its budget, and, most of all, its personnel. He must know their strengths and weaknesses and be both counsel and helper. He must have an excellent working relationship with his peers in organizations on which the project manager is dependent. In other words, he must be tapped into the grapevine and have a good action network to get things done should the project manager need help.

Consolidation of project reports to the highest level of management depends on how projects are treated within the organization. In some cases, they form a part of the budget or capital allocation of functional organizations or autonomous divisions of a corporation. In others, they form a part of separate capital appropriations at the corporate level. As a general rule, summarization of projects or their individual parts should stop at the first level after which they lose their identity.

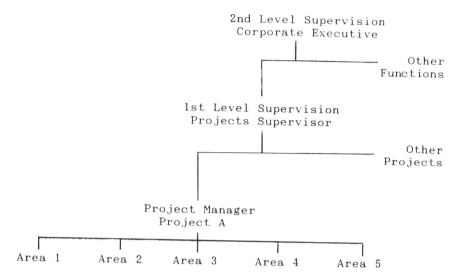

Figure 7.2   Project reporting summary levels.

Figure 7.2 illustrates this concept. At the first level, the projects are identified as A and B. Reporting to the second level, these projects would lose their identity and be combined for reporting above this level as the status and expectations for Capital Appropriation I. There are exceptions to this. Megaprojects are an example, as are the pet projects of corporate management, whatever their size.

It is recommended that the individual project managers participate in the preparation of all these reports. It is excellent training for determining what is important to management and reducing vast amounts of information to cogent summaries. The preparation involves those most familiar with the information and reduces or eliminates staff needs at the supervisory level.

7.2.4   Technical

Perhaps no project technical problem has received as wide attention as that of the booster rocket seal of the space shuttle Challenger. This was probably only one of the thousands of problems faced in this ambitious program. Unlike the many unknown others, it had two characteristics that make it typical of those which cause project failure. These are the failure to adequately solve the problem and the failure to assess the risks associated with

proceeding without a solution. Twenty-twenty hindsight will allow
recognition of these characteristics in nearly all similar, if not as
catastrophic, situations.

There are only two types of technical problems with which projects are faced: those which are anticipated and those which are
not. Typical examples of the first are those associated with projects that take existing technology and attempt to scale-up into
large sizes and capacities. The seal of the space shuttle Challenger booster rocket is one of this type. Of the latter, the pin
failure in the wheel spar of the McDonald Douglas DC-10 aircraft,
which caused a disastrous crash in Chicago, comes to mind. In
either case, the action is the same although the timing would necessarily be different.

Quality control problems may fall into both the above categories. Cost pressures may have resulted in the selection of a supplier or contractor whose quality performance may be questionable.
In this instance, problems may be anticipated and extra vigilance
would serve to surface them when they can be appropriately resolved. It is possible that such problems will occur without warning. In most cases, tracking occurrences will usually isolate them
to a single contractor or supplier. In the worst cases they will
result from a failure in the system.

Problems, particularly those of a technical nature, have several characteristics. One of these is the consequences of the lack
of a solution. Another is the time anticipated for a solution. An
adjunct to the latter is the development of an alternative which
would eliminate the problem. Not unlike a risk analysis, one needs
to assess the ultimate consequences and the probability of their
occurrence should the problem not be solved. As time passes without solution, the risks do not change, it is the actual acceptance
of those risks which ultimately arrives. Management must decide
on the acceptance of those risks and have in place a system that
will ensure a high degree of probability it will get to make a
choice.

When a problem is identified, there will come a time when a
decision will be required: that is, whether to proceed to implement the project, or to await the problem's solution. In the case
of the DC-10 problem, that point in time was immediate. Proceeding will have acceptable or unacceptable consequences. They may
be inevitable, but the question is whether or not they require involvement of management at the highest level. For example, the
problem may involve computer hardware which actually prevents
implementation of the project. Proceeding beyond the point the
project was to have been completed may involve only modest additional cost; therefore, proceeding would be an acceptable decision and need not involve the highest management levels.

The space shuttle Challenger presents a different situation, one that should form the basis of this special control. The technical deficiencies of the seal were known. They had been discovered early in the development and were being addressed. The eventual catastrophe stemmed from a failure to properly evaluate the risks of the technical problems and from proceeding without a solution. If a problem resolution control had been in place, it would have required a risk assessment immediately on problem discovery. That risk assessment could have only concluded that if the problem was not solved, catastrophic loss was probable. This is without question an unacceptable risk. Though there may have been time to solve it before the first shuttle launch, the highest level of management would have been made aware of the situation. The problem solution, or progress toward one, would have received regular and routine review. If you don't think corporate management is interested in all the problems that may bring on the investigative press or poor public relations, just ask them. Hardly a week goes by when some small disaster hasn't occurred that corporate management would like to have avoided or undo. A problem resolution control system will minimize these occurrences.

As soon as a problem is identified, a risk analysis is performed. The risk that will occur if the problem is not solved by the time the project is to be commissioned is categorized. Within the categories of acceptable or unacceptable, management must make a decision. A possible list is given in Section 5.6. Each risk is assigned a probability of occurrence and management determines at what level review will occur. For example, if loss of life or pollution were involved at any probability, the problem would involve corporate review. An expected time of solution is provided and progress reported. At each review, an updated assessment is made and the problem review level changed as circumstances change. The objective is to follow problem resolution at appropriate levels. Where risks are involved, those who must bear the consequences are given the opportunity to determine their acceptability.

7.3   THE HUMAN FACTOR

It is often impossible to completely eliminate differences, disagreements and competition, and it is questionable if it is advisable to do so. An organization without visible signs of these natural human traits is normally mediocre, overly conformist and a dull place to work in. [4]

F. L. Harrison

There is no absolute foolproof system because there are no foolproof people. Whether by accident or intent, human involvement has an influence on the outcome of all activity. Selection of educated, competent, experienced people only serves to reduce the probability of failure, not eliminate it. The availability of elaborate systems and the application of sophisticated controls have a similar effect. The combination of these ingredients constitutes the recipe for success or failure.

A sure way to invite failure is to assume that all project managers are alike and that systems will be equally applied. Good managers can overcome the weaknesses in poor systems and bad managers can cancel or negate the anticipated results of good systems. As each project is a unique effort, so too are its project manager and his application of whether systems are used. Ensuring project success, or analyzing project failure, begins with a careful examination of its ingredients and how they combine.

Table 7.4 provides a capsule of personal skills and characteristics and the consequences of weakness to which they may lead. Poor systems may compound these weaknesses and good systems may mask them. This would have some effect on the possible consequences, but will not eliminate them. It is unusual that these occur in isolation. One or all characteristics may be present simultaneously.

The primary responsibility for project execution is its manager, but first-line supervision bears a commensurate responsibility to recognize, overcome, and shore up weakness. Many project failures can be attributed to the abdication of this responsibility by first-line supervision and even higher-level management. Self-assessment is not common at any level. As one moves higher into the organization, appraisal by one's superior becomes a less objective exercise, if it is done at all. There is that all too familiar question, who audits the auditor? Prevention is better than cure. A mechanism available to provide a total overview is the independent project review.

There is much to be said for independent appraisal of project performance, but it is still only rarely done. Even when done, it must be well organized, staffed, and clear of purpose. It must be understood that success, failure, or accuracy in assessment of condition will be as much the responsibility of the reviewers as of the perpetrators. Observation must be broad based. It looks down into the project, to the systems under which it is functioning, and up to its supervision and management. The reviewers can be assembled from internal sources or an outside agency. Arguments can be made for both. Preventive reviews, those made with the objective of detecting early warning signs, are best handled internally as they can be done more efficiently,

Table 7.4   Consequences of Deficiencies in Project Manager Skills and Character

Skill or character deficiency	Adverse consequences
Technical	1. Inability to evaluate and judge the advice of others 2. Inability to referee technical disputes of subordinates 3. Overlooked opportunities for alternatives or discovery of hidden faults in scope 4. Lack of conviction and strength in presenting project to others (need to fall back on better-qualified subordinates)
Financial	1. Poor project evaluation 2. Consideration of fewer financial alternatives 3. Project overruns
Administrative	1. Confusion, disorganization, poor project records 2. Increased risks in claims exposure
People	1. Low morale 2. Poor communications 3. Lack of confidence by superiors, peers, and subordinates
Values	1. Lack of trust by all parties
Ambition High	1. Subordination of others' contribution to the project 2. Understating real problems and deferring help until too late
Low	1. Lack of confidence of subordinates 2. Failure to exert influence in progressing the project

inexpensively, and more often. For projects in trouble, an outside resource, such as consultants or other specialists, may result
in greater objectivity although more time consuming and expensive.

A useful review requires a systematic approach. The first
part of that approach is the objective of the review in the first
place. This is important in the failing project. It must be assumed that, if it is still failing after having been reviewed in the
past and not corrected, something was wrong that was not uncovered. As is often the case, the problem was recognized, but its
true cause was not and still persists. Patient, thorough investigation will enable it to be done right the first time and avoid having to do it over. The objective is to catalog the various aspects
that apply to a particular project. They will be assessed in the
review with personnel associated with the project and examination
of its records. Modifications, additions, and deletions are made
to suit the situation. Reviewers make their individual assessments,
compare notes, and utilize the results to determine the causes of
the failure.

Personnel qualifications should be checked to see how well they
match the job requirements. Turnover might be checked against
past experience on similar projects. Individuals can be interviewed
to assess morale and determine whether they understand their assignments and contribution to the project. Administratively, records should be checked for completeness, accessibility, and confidentiality. Reports should be timely and coverage complete. In-
place control systems must be understood by all personnel and
their effectiveness assessed. The budget and schedule must be
examined for reasonableness.

7.3.1   Special Situations

Two situations can result in project failure and in themselves are
failures. One is the case of the overzealous project manager and
the other is that of dishonesty.

The outward manifestations of success provide a means by
which personal progress and aspirations are measured. This can
be healthy ambition in a contained perspective, or it can be an
obsession that has no bounds. Webster uses such terms as "fol-
lower" and "disciple" as synonyms for "zealot." It is in this context in which they can be found in a project. If ignored or given
encouragement, they can be damaging.

To the zealots, the cause or objective becomes overriding. The
attitude is that the end justifies the means. Zealots become followers to their cause and will cut corners when necessary, utilize
the absence of control, or violate the rules to achieve perceived
ends. This characteristic manifests itself in behavior that makes

this type easy to identify. In order to confirm adherence to a cause, it is necessary to advertise disdain for the restraints which appear to detract from reaching the goal. There is also the need to brag about cleverness in beating the system. Absent constructive criticism, constant negativism is often ignored. In many instances, this is a lighted fuse. The ultimate explosion into a serious problem is only a short time off.

Combine the zealot with a weak system or a management that chooses to look the other way and you have a combination which often leads to disaster. Zealous managers may show up unexpectedly. With careful supervision and guidance they can be extricated with minimum damage. Their ability to get around the system often allows them to excel in emergency situations where systems are not in place to cope; but they leave considerable wreckage in their wake and require more maintenance than the apparent benefits would indicate. If the ability to cut corners is what makes them successful, perhaps the corners shouldn't be there in the first place so no one would have to cut them.

The overzealous individual fails to realize success may only be due to a management that chooses to look the other way. One misstep and it will all be over. Nowhere has this been more apparent than in the case of the Iranian arms deal uncovered in November 1986. Several minor players in the events met with an early demise. Only time will tell whether the real culprits will be exposed.

Misconduct by personnel on a project does not necessarily result in a failure of the project but certainly detracts from its success. Whether it is as serious as a criminal act, or as incidental as a conflict of interest, its revelation detracts from the objective at hand. Resources are diverted to uncover how it happened and restraints are imposed on the innocent. To deny that such things occur and to mire them in a cloak of secrecy only encourages their future perpetration.

The award of contracts, purchase of commodities and services, and their acceptance and certification provide ample opportunities for collusion, kickbacks, and outright theft. Neither procedures and safeguards to prevent their occurrence nor penalties for their commitment have been totally successful in eliminating them. On the other hand, recent statistics have indicated these so-called "white collar" crimes are increasing. How far one is prepared to go in attempting to prevent them depends on how often one has been "burned."

Several actions can help prevent misappropriation of funds, conflict of interest, and other undesirable conduct. These are:

1. Design and implementation of control systems covering all actions involving the receipt or disbursement of funds

and the recommendation, selection, receipt, or certification of commodities or services.

2. A statement of corporate principles and a code of conduct regarding what consitutes a conflict of interest. A written acknowledgment by each employee is recommended.
3. Unannounced audits of internal systems and procedures.
4. Prompt, decisive, and well-publicized punishment of offenders.

It is necessary to avoid creating an environment which, while enhancing the climate for better conduct, stifles action. Where this balance exists will be different for each organization. There will be a better chance to achieve a rational design if the controls are based on the concept that people are fundamentally honest. If one approaches control from the other extreme, some misconduct might be discouraged but action is slowed, opportunities missed, and morale suffers.

7.4 TURNAROUND OR FULL SPEED AHEAD?

When is a troubled project not a troubled project? When the problems are being isolated, their causes identified, solution developed and implemented. No project will proceed without trouble of some kind developing. It may be negative in the sense that there will be an impact on cost and schedule. It may be positive to the extent an opportunity has been lost to complete earlier or at lower cost. It is troubled when management does not recognize the problems or what to do about them.

Errors are not uncommon and, as previously discussed, are inevitable. Why is it when this predictable and inevitable event occurs, it causes a state of panic? True, in some cases the error is beyond the desirable limit, but the laws of probability do not preclude the occurrence of extreme events. This does not mean one should sweep incompetence, or indiscretion, under the rug. There are, however, alternatives to throwing project managers to the wolves, demoralizing what is otherwise an effective project team, and creating trouble where none need exist.

Before any action is taken, an assessment of the real problems must be made. Examined in isolation, they may not be as severe as first assumed. It is possible that they can be adequately treated and the project continued without major surgery. An outside task force may be able to attack the problem, while the rest of the project moves on in parallel.

These are narrow limits in which this technique can be effective, but they do occur often enough to take a hard look at the situation and see if they apply. The condition of potential cost

overrun, schedule delay, or both could have resulted from any or
a combination of the following:

1.   Events beyond the reasonable control of the project's man-
      agement (force majeure)
2.   The consequences of risks initially accepted

The condition should not have resulted from similar events which
had previously occurred or actions previously taken. Neither
should it have resulted from the incompetence or inexperience of
the project's management. If the criteria are met, a revised estimate of cost and a new schedule can be developed. If the justification still exists to proceed with the project, the slate is wiped
clean. The new basis becomes that by which progress and performance are to be measured.

7.5   THEY ARE EXPENDABLE

Contrary to naval tradition, corporate captains do not
go down with the project ship; they jump long before
the sailors even suspect rising water. The best way to
tell if a project is doomed to failure is not to watch
for the rats abandoning ship but to keep your eye on
the fat cats instead. They give a much earlier and
pronounced warning. [1, p.13]

                                          Robert D. Gilbreath

The man at the head of the column may take pride in the fact
he has been chosen as leader. More often than not, when things
go wrong, he is also the first casualty. Parallels often seem to be
made to professional sports teams where managers certainly have
a shaky tenure. The parallel with the business manager is incor-
rect. In the sports context, the manager is not the star, it is
the players who are stars. The manager's job is to blend the individual stars into a strong team. On a project, it is the project
manager who is the star. He has been selected for his ability,
whether proven or assumed, to take an average group of supporting players and blend them into a strong team. This difference
is important. Just look around and see how many big league managers have rebounded from losers to winners. Leo Durocher in
baseball and Paul Brown in football are two memorable men who
come to mind. Then ask how many project managers got a second
chance after managing a troubled or failed project. Their stars fade
in the pall that surrounds the problems they create or fail to solve.
They are expendable.

Misconduct presents a clear case for dismissal of a project manager. Under such conditions the action should be decisive and promptly implemented. Irrespective of the immediate disruption that may occur, a clear message that misconduct is not acceptable and will not be tolerated is more important. To be effective, the reasons for the dismissal should be officially communicated, even if by word of mouth. If the company is not prepared to go beyond dismissal, the case should be well supported in the event of recrimination.

Unfortunately, most of the conditions that lead to dismissal of the project's manager are not always clear cut. As it is an irreversible choice, it must be made with all due deliberation and when little else is likely to bring desired results. The question to be answered is: Are the personal weaknesses of the manager so great that little else will result in halting the slide or turn the situation around? The answer is never straightforward. In the back of management's mind is a nagging question. If the project manager allowed this situation to develop and cannot correct it, how did he get to be in charge in the first place?

There is a risk that management, in denial of its own contribution to a troubled project, will fail to take decisive action. The project will continue to flounder, morale will sink lower, and the cost to stem the rising tide of cost and schedule slippage will continue unabated. Enlightened managements are rare. They do exist and, because they are, don't often find themselves in the position of having to extricate themselves from projects in such trouble that a project manager's dismissal is the only alternative.

7.6   THE ULTIMATE STEP

The Spanish have called it "the moment of truth," in reference to facing the final charge of the bull in preparation to planting the sword. Others might liken it to "being caught between a rock and a hard place." Whatever you call it, having to decide to shut down a project is a difficult, often painful decision. For some projects it is not only the merciful, but the necessary, thing to do. How and when is this conclusion reached? There are so many variables and so little applicable precedent that each case must be decided on its own. But some common factors can be considered and be of benefit in making a decision.

Some projects are marginal at the time they are conceived. It is possible that they were only slightly better than some alternatives. In other cases, they may be viable under a given set of circumstances, but subject to change if circumstances differ.

At their inception, a time should be established at which their continuation is brought up for confirmation. The appropriate time is when the cost of possible abandonment is low. It may only be possible, however, when their justification can be anticipated to change. In either case, it should be a point of choice, rather than chance.

In most instances, the decision to continue or abandon a project is the result of the accumulation of serious difficulties. The decision must be a conscious one, based on an evaluation and assessment of all of the factors that influence that choice. They may not all be purely economic. An incomplete list of these is given below:

1. Profitability of project if continued
2. Tax consequences of spent funds
3. Public, stockholder, employee, or union relations
4. Cancellation costs of commitments
5. Salvage value of unwanted assets
6. Contractor and supplier goodwill
7. Alternative uses of unspent budgeted funds
8. Disposition of project personnel

Frustration, disillusion, and even fear attend a project in trouble deep enough to consider its abandonment. These conditions require a fair evaluation and the making of the right decision. No matter which direction the decision, it will be questioned. Emotional climate notwithstanding, the facts must be carefully sifted and assessments carefully weighed. If the choice is to abandon the project, it must be done swiftly and with dispatch.

7.7   SUMMARY

There is no panacea for rescuing a project in trouble. The problems are never simple and are usually the result of a combination of factors. Solution begins with a correct diagnosis of the problems and their causes. The lack of any ready formula portends difficult and painstaking investigation. The effort is enhanced by comparing what has been done with what should have been done.

Prevention is still the single most effective mechanism for averting project trouble. An experienced and skilled project manager coupled with cogent controls is a first line of that prevention. Timely in-depth reviews aid in detecting problems at an early stage.

The interdependent nature of activities on a project demands that a search for problem causes include all aspects of project execution. This means a thorough analysis of cost control, scheduling, and adminstrative practices in accordance with established procedures. The procedures themselves are subject to examination

in view of what is required. Costs are examined in comparison
to the original budget estimates. Trends in related cost elements
can reveal the cause of current problems, or the potential for future ones. The handling of contingent funds can mask otherwise
obvious areas where controls have failed or are needed. Schedules, like costs, become better known as the project progresses.
What is often replaces what should be, and manipulation of event
sequences results in reduction of optimal alternatives. Comparison
with original critical events will lead to the reasons activity float
or slack time seems to drain away until nearly all events are critical. Lack of accurate, well-organized records is a feature of
projects in trouble. What can assist in diagnosing the problem is
not there. This is taken as an early warning sign of potential
trouble, even where none yet exists. Reporting is a key element
of project administration. The information presented should be
timely, accurate, and reflective of the responsibilities of the levels
of project management who initiate it.

Projects are managed by people. Very few projects have the
benefit of managers with broad experience required to run them.
It is necessary that a project's manager recognize the areas of
weakness and attenuate them with strong subordinates. It is management's responsibility to know its managers and their capabilities.
Management should build on the strengths and support the weaknesses of its managers to the extent the project team is well balanced. There are times when it is desirable, if not necessary, to
replace a project's manager. This action should be carefully considered.

The chaos, confusion, and finger pointing attendant upon a
project in trouble often hide the fact that the cause may have
been beyond the reasonable control of the project's management.
In this event, the cause must be confirmed and, if the project is
still viable, the slate wiped clean. A new schedule and budget
can be established and form the basis on which the project will
then be measured.

There are times when the problems are so severe as to warrant
consideration of abandoning the project. The consequences of such
a decision must be carefully and accurately weighed. If the result
is demise of the project, it must be terminated as expendiently as
possible.

EPILOGUE

It is difficult to say which is more rewarding: completing a project successfully where everything went smoothly and according
to plan, or salvaging a project where a lot of things went wrong.

In either case, the defined objective and temporary nature of projects provide the opportunity to experience the full range of physical and emotional involvement in a productive effort in a short period of time. Be it a system, a process, or a physical structure, a project has an identity for which all of the participants can share credit.

Managing effectively cannot be taught. What can be taught, however, are the tools and techniques of management. Experience itself becomes the teacher of when these tools and techniques are appropriate for application. My attempt here has been to identify some of these which have served well over a long period in a variety of circumstances. In addition, there has been the caution that there are also pitfalls. I have found that the best project managers are those who couple optimism with realism. They must be optimistic to the extent they provide direction and encouragement to the team to succeed. They must be realistic, so as not to overlook the fact that problems will occur, no matter how well planned the effort. Finally, a good project manager must understand people. In most cases, persuasion becomes the only method by which the objective will be achieved. The ability to know not only when to give and take, but how, is the most important skill in his arsenal.

REFERENCES

1. Robert D. Gilbreath, *Winning at Project Management*, John Wiley & Sons, New York, 1986.
2. Robert J. Graham, *Project Management; Combining Technical and Behavioral Approaches for Effective Implementation*, Van Nostrand Reinhold, New York, 1985, p. 30.
3. M. C. Grool, J. Visser, W. J. Vriethoff, and G. Wijnen, eds., *Project Management in Progress, Tools and Strategies for the 90's*, North Holland, Amsterdam, 1986, p. 32.
4. F. L. Harrison, *Advanced Project Management*, John Wiley & Sons, New York, 1981, p. 309.

Appendix: Contractor Evaluation

The following contractor evaluation worksheet is not intended to
be complete and should be expanded as necessary to accommodate
the type of contractor, type of contract, and any special considerations peculiar to the project on which the contractor will perform. It is intended to be general in nature and many of its features will apply to all contractors, although some may be specific
to consultants, design/engineering contractors, service contractors,
or construction contractors. The importance of personnel evaluation should not be overlooked. It may not be possible to control
the assignment of personnel on a lump-sum arrangement, but regardless of contract type, the people assigned to the project will
have the most significant impact on its success or failure.

Parenthetical comments have been appended to many of the
evaluation items in order to elaborate on their purpose or importance. It is recommended that in the development of specific evaluation forms, evaluators be apprised of the intent of the forms,
the objective of the evaluation, and be briefed. The purpose of
the briefing is to assure that each evaluator understands the basis
for each evaluation item. Because the importance of individual
items will vary with the type of contract and contractor, it is
recommended that weightings be assigned. The evaluators should
agree on the weightings, particularly if the evaluation is to be
used to assign financial adjustments to the contract price of anticipated contract cost.

Experience indicates that evaluators should have varied backgrounds and be in a position so as not to be unduly influenced
by one of their number. The evaluations should be made independently and then reviewed with the entire group in order to

moderate extremes. The goal is to achieve consensus, but if this
is not possible, someone should be in a position to artibrate the
disagreement or render a final decision.

CONTRACTOR EVALUATION WORKSHEET

Organization

Corporate

  Form

    Centralized

      Functional_ _
      Industry_ _

    Divisionalized

      Geographic_ _
      Functional_ _
      Industry_ _

    Experience

      General _
      Industry_ _
      Specific project type_ _

    The corporate form of organization determines how the project
will be administered within the organization and the access the project has to the key decision makers. The centralized form can ensure more ready access, but if there are many projects, decision
makers may be overloaded and their interests spread too thin. A
judgment based on work load is necessary.

    If the organization is functionally structured, it is highly likely
that the project administration will be also. This means a matrix
operational mode and the concerns expressed in Chapter 3 must be
addressed. Interviews with key project personnel will determine
the nature of the decision-making apparatus. Indications that the
project manager's decisions are subject to review by functional
management can be considered a serious negative.

    The divisionalized form can provide a delegation of authority which
places decision makers close to the project or adds a layer of bureaucracy. One method to ascertain the efficiency of this form is
to interview other clients on current or recently completed projects.

    A geographic division may mean that support to the project may
be close at hand or may have to be relayed to the home office.

If the division handles numerous similar projects, it may have all the expertise necessary.

Industry divisionalization is most common in construction and engineering due to the value of experience with specific industry applications and the client relationships. This ensures that similar projects are being executed and expertise will be related.

Corporate organization experience is a measure of the exposure to the technology. It is an inadequate measure in that time cannot be equated directly with knowledge, but serves as the only readily available proxy. Assuming that corporate staff and management can be called on by the project for guidance and assistance, the cumulative experience of those employed in applicable project areas represents a comparative guide between contractors.

Workload

Current versus capacity _
Growth rate _
Percent of capacity – this project _ _ _ _ _ _ _ _ _ _ _ _ _ _ _ _ _ _

A contractor's workload provides significant clues as to how important your project is and how much attention it might get in the face of current and future demands. If current work is well below capacity, it may mean a better fee structure to obtain new work. It may also mean that the contractor is beginning to shift the people he wishes to retain to current work and all who are available will be those he will surplus if he cannot obtain work. Interviews can determine what jobs personnel were working on and the status of those jobs. Seniority of those to be assigned to your job versus the company average is another clue.

Growth is the objective of any enterprise. If this growth is very rapid, it can mean that experienced personnel are being spread thin and many of the newcomers to an organization are still on the learning curve. To a client, this can be expensive education. It also leads to reorganizations in order to manage the growing business and the trauma associated with it. Growth much beyond that experienced by the industry in general and the project type in particular should be approached with caution.

The percentage of the total contractor capacity occupied by a single project affects the growth and attention rate. The lower the percentage, the less important any one particular project is to the contractor. However, it also has a lesser impact on the contractor's resources and enhances his ability to draw on those resources to staff any given project or react to unforseen circumstances. One rule of thumb is that a single project should not

consume more than 20% of a contractor's resources in any one location. This must be tempered by a careful analysis of the project's demands and the contractor's over time as provided by the
loading diagrams. A sample is given in Figure 4.2.

Project

    Task force_ _
    Functional_ _
    Matrix_ _

The project organization reflects corporate attitudes toward
project execution. Project-driven organizations will almost always
establish autonomous teams in a matrix form from resources which
develop a loyalty to the project. Task forces are normally reserved for short-term projects which involve several disciplines
where expertise or involvement is determined beneficial. Functional organizations are adaptable to projects where one discipline or
beneficiary predominates.
    Evaluating a contractor's project organization within these forms
depends on two factors, autonomy and dedication. The first relates to the capacity of the project organization to execute the project with its own resources or its dependance on others for support. On a large project, for example, the project organization
may be completely autonomous in all aspects except for purchasing, which may be centralized. There are those who claim the
measurable cost differences of dedicated versus centralized procurement are offset by lower hidden costs. Unfortunately, there
are no simple guides to aid in this decision. Dedication is measured by the full- or part-time assignment of personnel to the project organization. Like a man with two bosses, there are times
when one must choose which to serve. There are no objective
guides to this choice.
    Note: As explained in the text, the project organization ulti-
mately becomes a marriage of the client and contractor organizations. If this is to be a successful marriage, it depends on the
openness of both parties as to their intentions and operational
modes. It is strongly recommended that in the tender inquiry to
contractors, the client clearly indicate the form of organization for
his project team and the manner in which the contract is to be
administered. This includes the authorities of the project manager and subordinates and how they will interface with the contractor's personnel.

Personnel

Corporate sponsor/Project manager/Key subordinates

General experience_ _
Project administration experience_ _ _ _ _ _ _ _ _ _ _ _ _ _ _ _ _
Years with present employer_ _ _ _ _ _ _ _ _ _ _ _ _ _ _ _ _ _ _
Number of employers, including present _ _ _ _ _ _ _ _ _ _ _ _ _
Cumulative years of experience with team _ _ _ _ _ _ _ _ _ _ _ _

The most important aspect of contractor performance is the quality and ability of its personnel. This is true irrespective of the
type of contractor or contract. Besides capability, the attitude a
client should look for in contractor personnel is that doing a good
job for the client will result in doing a good job for the employer.

General experience indicates the broadness of knowledge and the
potential of tapping a multiplicity of sources for ideas and solutions. Administrative and project management experience indicates
the capability to direct the efforts of others and employ various
systems in the prosecution of work.

Years of experience with the current employer indicates the exposure to its systems and procedures. It also measures the opportunity to develop informal networks and become part of the information network. There is no measurement for the value of this experience, but one would intuitively expect that those with greater
opportunity to develop these networks have means to hasten action, get the ear of key personnel, and disseminate information
rapidly.

Project management is a team effort. It is also a temporary effort. This implies that new teams must learn to work with each
other at the same time they are expected to produce results. If
team members have already worked together, they should know
each other's capabilities, weaknesses, and personalities. The
shorter the project, the greater the value of these previous associations.

Although working for several employers provides an individual
with varying perspectives in the approach to project execution,
having too many employers carries the stigma of instability. What
that number is can vary, but anything over three should prompt
careful assessment. By the time the key members of the contractor's team reach 10 to 15 years' experience, they should be considered on track to attain project manager status, or higher, within the organization. The higher they are in the project organization, the longer should be their experience with their current

employer. It should be remembered that for many in the project
management business, achieving project manager status is the ultimate goal. Attitudes and objectives of the key individuals can
be ascertained during personal interviews.

Systems

Information

Procedures manuals _
Library _
Management information systems (MIS) _ _ _ _ _ _ _ _ _ _ _ _ _ _

Procurement

Open orders_ _
Policies _
Procedures _

Scheduling

Procedures _
Flexibility_ _

Cost Control

Procedures _
Flexibility_ _

Engineering/design

Procedures _
Computer-aided design_ _
Expert systems _

Procedural formality runs the gamut from nonexistent to cumbersome and seldom used. The important factor is consistency. This
can be accomplished with minimum formality and limited written
documentation. The more documentation, the greater the likeli-
hood that few have ever read it. How a contractor administers
his work on one project should be essentially the same as on another. The objective is not to have to climb the learning curve
on each new project. All of the contractor's personnel should be
familiar with the systems applicable to their area, know how to use
them, and how to apply them consistently on each project.
    Management information systems have come into vogue and can
provide another useful tool in a contractor's arsenal. These too
can become overly sophisticated and fall into disuse. The objective of any information is to provide those who need to know
the information required to perform their jobs. Regardless of
the system used, personal interviews of contractor personnel,

including its executives, can ascertain whether they are in touch
with the elements of their function. If they volunteer the types
of data that pass across their desks, it is a healthy sign they feel
they are on top of their jobs. Conversely, they should be queried as to how often and when was the last time they had direct
contact with the projects for which they are responsible and the
people who run them.

Many contractors maintain elaborate libraries of reference material for their operations. These incorporate standards, codes, materials catalogs, supplier and vendor catalogs, and even training
materials. The important thing is not their presence, but their
usage. A quick check with the librarian in these facilities will
ascertain how often they are used and who uses them.

The larger the contractor, the higher the volume of procurement
activity. This higher volume can translate into significant savings,
which can be passed on to the client in the form of lower prices
for service. In addition, large volume buyers will get preference
from vendors in deliveries and when supplies are in short supply.
Some contractors have achieved these benefits and still maintained
project control of purchasing. This is accomplished through open
orders which can be used by the project procurement personnel
without having to use the services of a centralized buying office.
This eliminates the middleman and places the project in direct control of the order with the supplier. In this respect it is important
to determine how the contractor resolves conflicts between projects
when the supplier cannot satisfy both demands.

Cost and schedule systems should be flexible to allow the contractor to operate in the manner in which he is most familiar, yet
adaptable to the needs of the client. In many cases, the needs
are similar and the contractor will normally utilize a more detailed
breakdown than that required for summary reporting. Often, the
only thing required may be a conversion program which uses contractor data to generate client reports.

Lump-sum contracting prevents access to complete contractor
cost data. This should not discourage monitoring of the contractor's effort. A client who does not attempt to keep track of a
contractor's financial activities on a project has only himself to
blame when the contractor gets into financial difficulty to the extent the project begins to suffer. Approximate costs of labor,
materials, and cash flow are no big secret. Engineering estimates
prepared during budgeting and design should provide an adequate
mirror for contractor costs. Maintaining cash flow accounts during
project execution is low-cost insurance and gives an early warning
of impending problems which may affect project progress.

The key to scheduling systems is the level at which they are
used. A simple bar chart, where development and execution reach

down to the individual craftsman, is far superior to an elaborate
precedence network, which becomes an exercise between schedulers.
In addition to personal interviews of contractor schedulers and planners, it is important to query former clients as to this aspect of a
contractor's system. Even after the project starts, it is not too
late to casually ask the foreman what he plans to do the next day.

Computer-aided design (CAD) was only a toy several years ago.
Today it has become a sophisticated tool of many uses. Expert
systems are in the same position today as CAD was then, but have
an even greater potential. The caution remains that few people are
intimately familiar with the mechanics of either and are easily influenced by the fancy graphic displays which can be created and
the claims for expert systems. Before opting for any of these systems it is important to pin down in writing: what is to be used,
what is to be accomplished, how much it is to cost, and a commitment that if the results are not achieved, there will be no charge.
This condition should temper both the claims and the expense.

The answers to most of the questions in a contractor evaluation
should be solicited through development of the tender request.
This obtains the answers in writing and, through the contract, can
be incorporated as legally binding. This is no substitute for the
personal contact made possible by personal interviews. As indicated, these interviews should include all of the key project personnel, the project sponsor, and corporate executives. Selective
interviews and queries of support staff units is appropriate if they
are also to be involved in the project. The contractor should be
asked to provide a list of his most recent projects, his clients, and
the names of contacts with those clients. If at all possible, client
project managers should be contacted to provide additional background.

It is impossible to develop a detailed evaluation form that would
be applicable to all types of contractors and contracts. The foregoing is intended to outline the areas that will be encountered in
most situations and provide a guide to the user enabling the development of the form for specific application. Following the above
format, a weighting would first be assigned to the three primary
elements: organization, personnel, and systems. These weights
would be distributed among the items under each of these elements.
A scoring system would be established, perhaps a 1-to-5 scale progressing from poor to below average, average, above average, and
outstanding. The final score would result from multiplying the
score times the weight. An example is given below.

Personnel	Weight	Score	W × S
Project manager — John Smith	0.20	—	
a. General experience	0.03	3	0.09
b. Project administration experience	0.03	4	0.12
c. Years with current employer	0.04	3	0.12
d. Number of employers	0.04	4	0.16
e. Cumulative years with team	0.06	5	0.30

Project — Specific criteria

	Years				Years
	a	b	c	d	e^a
1.	0 − 5	0 − 3	0 − 3	>5 or <2	0 − 5
2.	5 − 8	3 − 5	3 − 5	5	5 − 8
3.	8 − 12	5 − 8	5 − 8	4	8 − 10
4.	12 − 15	8 − 10	8 − 10	3	10 − 12
5.	>15	>10	>10	2	>12

[a] Assumes five key subordinates: engineering, procurement, construction, administration, and controls.

Index